RESEARCH INTEGRATION USING DIALOGUE METHODS

RESEARCH INTEGRATION USING DIALOGUE METHODS

DAVID MCDONALD
GABRIELE BAMMER
PETER DEANE

ANU

THE AUSTRALIAN NATIONAL UNIVERSITY

E PRESS

ANU
E PRESS

Published by ANU E Press
The Australian National University
Canberra ACT 0200, Australia
Email: anuepress@anu.edu.au
This title is also available online at: http://epress.anu.edu.au/dialogue_methods_
citation.html

National Library of Australia
Cataloguing-in-Publication entry

Author: McDonald, David.

Title: Research integration using dialogue methods / David
                McDonald, Gabriele Bammer, Peter Deane.

ISBN: 9781921536748 (pbk.) 9781921536755 (pdf)

Notes: Bibliography.

Subjects: Interdisciplinary approach to knowledge.
                Research--Methodology.

Other Authors/Contributors:
                Bammer, Gabriele.
                Deane, Peter.

Dewey Number: 001.4

Cover design by ANU E Press

Table of Contents

List of Tables

Acknowledgments and author contributions

Australian Government
Land & Water Australia

DPMP
Drug Policy Modelling Program

A Land & Water Australia Innovation Grant (ANU58) provided the main funding for the research underpinning this book. *Disclaimer:* By funding this research, Land & Water Australia supports gathering and disseminating information, not rendering professional advice or services. Land & Water Australia expressly disclaims liability to any person or organisation in respect of anything done or omitted to be done that is based on the whole or any part of this document. The Colonial Foundation Trust through the Drug Policy Modelling Program provided additional funding for searching out cases. A substantial amount of Gabriele Bammer's contribution to the writing of this book was undertaken while she was a Visiting Scholar at the Competence Centre Environment and Sustainability, Swiss Federal Institute of Technology, Zurich, in April and May 2007.

During the development of this project, we benefited from suggestions and critical insights provided by Lorrae van Kerkhoff and Alice Roughley. Valuable comments on a first draft were received from Lorrae and Alice, as well as from Wendy Gregory, Carolyn Hendriks, Gerald Midgley, Christian Pohl, Alison Ritter and Dan Walker. Caryn Anderson provided valuable comments on a later draft and on the index.

David McDonald searched out the dialogue tools and wrote the bulk of the descriptions and case studies. Gabriele Bammer conceptualised the study, teased out the elements of research integration and wrote the introductory and most of the concluding sections. Peter Deane hunted out case studies and sifted through what he found to develop a short list, which he and David then selected from. All three were involved in organising and refining the book. All links to electronic material were checked at time of publication.

Chapter 1. Introduction

Research integration is the process of improving the understanding of real-world problems by synthesising relevant knowledge from diverse disciplines and stakeholders. Methods for undertaking research integration have not, however, been well developed or explained. Here, we show how 14 methods developed for dialogue can be useful for research integration. What makes this book unique is that we tease apart components of research integration and match them to particular methods.

Research integration is essential for effectively investigating real-world problems. Such investigation requires bringing together the insights of different disciplines. For example, examination of the impacts of the encroachment of housing on farm and bushland on the fringes of cities can benefit from the expertise of ecologists, economists, hydrologists, sociologists, soil scientists, demographers and so on. Similarly, to comprehensively model the impact of the covert release of an infectious disease agent on a major city requires input from, among others, communicable disease epidemiologists, statistical modellers, urban geographers, psychologists and legal experts.

Bringing together such different disciplinary insights to more thoroughly understand a particular real-world problem requires a new type of researcher with a specific set of concepts and method skills. These skills complement those of disciplinary experts. One of us (Gabriele Bammer) has developed the foundations for a new cross-cutting discipline—Integration and Implementation Sciences (I2S)—which is designed to equip this new breed of researchers with the theory and methods necessary to provide effective integration in all forms of cross-disciplinary research, be they multi, inter or transdisciplinary. One of the essential skills is competence with various integration methods, including methods based on dialogue, as we describe here. We outline the full characteristics of Integration and Implementation Sciences in Appendix 1.

Research integration has another component, in addition to being able to pull together knowledge from the disciplines relevant to understanding a particular problem. This relates to recognition that, while academic disciplines provide essential knowledge about particular aspects of an issue, relevant knowledge is also held by various stakeholders, who are usually those affected by the particular problem and those in a position to make decisions about it. An integration researcher therefore also has to be skilled at involving those groups and weaving their insights into the composite understanding. In the case of land use in peri-urban areas, affected parties can include farmers and recreational users of bushland (whose activities are impinged on by the expansion of housing) and families requiring housing. Those in a position to make decisions about the issue include government policy-makers, regulators and land developers. In the case of modelling the impact of a disease release in a major city, a wide range of stakeholders would be involved in dealing with such a terrorist attack including police, emergency medicine specialists, media, politicians, and business and community leaders. An integration researcher needs to be equipped with the skills to capture the valuable expertise and insights of these groups to provide a more comprehensive understanding of the problem at stake.

Research integration therefore improves understanding of real-world problems by synthesising knowledge from relevant disciplines and stakeholders; it is integration in the context of research, integration by researchers and integration as a research activity in its own right. Research integration involves more than just bringing together knowledge in terms of 'facts'. It requires appreciation of different epistemologies (that is, the variety of different ways in which we can come to know 'something'), as well as different underlying values, interests, world views, and so on. A more comprehensive understanding of real-world problems involves teasing out such differences and finding ways to synthesise them. The role of a research integrator is to identify, gather, combine and analyse relevant disciplinary and stakeholder knowledge in a way that clarifies the diverse aspects of a problem, as well as the relationships and interconnections among them. The aim is to contribute to a more comprehensive understanding of the problem. This is useful in its own right, as well as contributing to the ability of decision makers to more effectively tackle the problem.

One key set of methods for gathering and combining such diverse perspectives builds on various techniques of dialogue. What do we mean by dialogue? We use Franco's (2006:814) definition, which is to 'jointly create meaning and shared understanding' through conversation. Of course, not all dialogue requires a 'method'. Dialogue can occur through the normal give and take of talking and listening, especially when two, or a small group of, people are involved. Once the group starts to become larger, however, structuring the conversation becomes essential for different understandings to be effectively shared and brought together. Many methods for organising conversation have been developed, but they are not all dialogue (debate is an example) or relevant to research integration. We have chosen 14 of the best-described methods to present here. The aim is to provide a methodological 'tool kit' to assist integration researchers in bringing together multiple perspectives—from disciplines and stakeholders—to address real-world problems. This is not a book about dialogue per se, although we provide a limited amount of additional background information on how these dialogue methods for research integration sit within the broader field of dialogue in Appendix 2.

While few researchers would, at this stage, describe themselves as integration researchers, let alone integration and implementation scientists, many have taken on an integrative role in cross-disciplinary research addressing real-world problems. They are our primary target audience. They are likely to already have some familiarity with at least some of these dialogue methods, but will be interested in expanding their repertoire. We want to provide them with not only a wider array of methods, but also the foundations on which to continue to develop dialogue methods for research integration.

A second audience is those who seek to become research integrators who are new to methods based on dialogue. For this group, the book aims to provide an overview of available methods that they can seek training and guidance in. Discipline and stakeholder experts who are invited to participate in research integration using dialogue methods are a third audience. We anticipate that improving their understanding about each particular method and what it is most suited to accomplish will enhance the success of the integration endeavour.

In producing this book, our approach was iterative and switched between inductive and deductive. We identified elements of research integration, such as synthesising facts, judgments, visions, values, interests, epistemologies, time scales, geographical scales and world views. At the same time, we read about dialogue methods, examining them through the lens of research integration. We cycled between identifying elements, different types of dialogue methods and case studies in order to match methods with integration tasks. We provide a more detailed description of our methods in Appendix 3.

We wanted to get a sense of the array of available methods, to explore how well they linked with specific research integration tasks and to present those that were the best described (rather than attempting to cover all dialogue methods). As far as we are aware, this has not been done previously. Furthermore, we wanted to not only link individual dialogue methods to specific elements of research integration, we wanted to provide examples of how the methods have been used in research integration. We looked for examples in four research areas: the environment, especially natural resource management, public health, security, and technological innovation. The aim here was twofold. First, we wanted to illustrate how these dialogue methods were broadly useful in a range of different areas. Second, new dialogue methods are often produced in relation to one area of application—for example, they can be produced by researchers investigating environmental problems—and there is no accepted institutional pathway for researchers studying problems in other areas, such as public health, to become familiar with such innovations. This book aims to enhance possibilities for cross-fertilisation.

So far, we have explained what we mean by research integration and by dialogue methods, as well as presenting a brief overview of our approach. In the next section, we discuss how we identify and classify the dialogue methods presented. Following that, we examine a range of issues to do with the application of these dialogue methods. The bulk of the book consists of descriptions of 14 dialogue methods for research integration, along with case studies illustrating how they have been used for this purpose. The penultimate section of the book describes an exercise matching the methods to the sorts of dialogue questions they are most suited for, using a hypothetical example. This further differentiates methods, allowing research integrators to choose the technique that is most appropriate for a specific research integration task. In the conclusion, we outline how the use of dialogue methods in research integration can be enhanced—particularly by better documentation and publishing of successful and unsuccessful cases, by developing new methods, by continued cross-fertilisation across different topic areas and by improved critical analysis and evaluation.

This book charts new territory in linking dialogue methods to specific research integration tasks, and also provides the foundations for further development of dialogue methods for research integration. We believe that this is a fertile field, which will contribute better solutions to the complex problems facing society.

Chapter 2. Using the dialogue methods in this book

Identifying and classifying the dialogue methods

In this section, we provide an overview of the dialogue methods, describing how we classify them. It is useful to reiterate that our aim is to present group conversation processes to jointly create meaning and shared understanding about real-world problems by bringing together knowledge from relevant disciplines and stakeholders.

The challenging issue for identifying relevant dialogue methods and classifying them relates to just what is being integrated. From an initial understanding of what (structured) dialogue might integrate, we developed a list of elements we believed were possibly being integrated—these included facts, judgments, visions, values, interests, epistemologies, time scales, geographical scales and world views. These provided the basis for further interrogating the literature on dialogue and hunting out case studies. In terms of the elements we identified, we found dialogue methods specifically geared to integrating judgments, visions, world views, interests and values.

In this way, we determined that there were two broad classes of dialogue methods for research integration: those that were useful for gaining a broad understanding of a problem and those that were useful for honing in on a particular aspect of a problem.

We put methods for integrating judgments together to make up the class of methods for gaining a broad understanding. In forming a judgment, a person takes into account the facts as they understand them, their personal goals and moral values, and their sense of what is best for others as well as themselves (Yankelovich 1999). Most of the dialogue methods we identified fell into this group and they are citizens' jury, consensus conference, consensus development panel, Delphi technique, future search conference, most significant change technique, nominal group technique, open space technology, scenario planning and soft systems methodology.

The second class of methods focuses on a particular aspect of understanding a problem. We identified methods specifically geared to four aspects: integrating visions (appreciative inquiry), world views (strategic assumption surfacing and testing), interests (principled negotiation) and values (ethical matrix).

Before moving on to describe these groups of methods, it might be useful to outline how we think they could be used for research integration, or more particularly how we think they should not be used. We do not believe that research integration needs to slavishly identify every element of knowledge and then institute a process for bringing together all the disciplinary and stakeholder perspectives on each element. Instead, for most problems, a method for developing broad, shared understanding, as indentified in our first class of methods, will be more than adequate. For some problems, however, it can be particularly important to tease out one aspect. For example, in the peri-urban land-use illustration, understanding different values about progress and growth, conserving the environment and providing equity for all citizens (in terms of access to housing, in this case) will be integral to developing shared understanding, so that a dialogue method targeted at values can be particularly helpful.

Similarly, for other problems, differences in visions can be particularly pertinent. Vision here relates to aspirations about dealing with the problem. For example, if the problem under investigation is the different life expectancy between rich and poor members of a community, different ultimate aspirations can affect the ability to bring different perspectives together. Those whose vision is to use the community as a case study to develop national policy tackling multiple facets of poverty will approach the problem differently from those whose aspiration is to improve employment opportunities for the disadvantaged in that one area. When the problem is such that the disciplinary and stakeholder experts are likely to have widely different visions, methods focusing on understanding these could be necessary.

The same logic applies to world views or mental models, which are the assumptions that each of us hold about how the world works in relation to the problem under consideration. That logic also applies to interests, which are our motivations for getting involved in understanding the problem.

We therefore classified the methods we identified as useful for research integration as follows.

I. Dialogue methods for understanding a problem broadly: integrating judgments:

- citizens' jury
- consensus conference
- consensus development panel
- Delphi technique
- future search conference
- most significant change technique
- nominal group technique
- open space technology
- scenario planning
- soft systems methodology.

II. Dialogue methods for understanding particular aspects of a problem: integrating visions, world views, interests and values:

- appreciative inquiry: integrating visions
- strategic assumption surfacing and testing: integrating world views
- principled negotiation: integrating interests
- ethical matrix: integrating values.

As with all classifications, the boundaries between different groups are not hard and fast. This is compounded further by the flexibility with which particular methods can be applied. Nevertheless, we suggest that the classification we present here provides a workable beginning that can be used as the basis for further development of dialogue methods for research integration.

Before moving on to issues concerning the application of these methods, it is also important to point out that, by and large, the dialogue methods we investigated were devised for some purpose other than research integration. For many, it is an easy, logical move to increase their applications to include research integration. For some, however, expanding their use to research integration requires a different way of thinking about the method. For example, the nominal group technique falls into the former category. This is a highly structured method to assist participants in pooling their judgments about an issue, involving the generation, recording and discussion of, and voting on, ideas. As we illustrate in the relevant section of this book, there are clear examples of how this is useful in research integration. On the other hand, using principled negotiation for research integration requires thinking about this method in a novel way. Principled negotiation was originally devised as a conflict-resolution method but its techniques—for identifying interests, generating options for meeting the range of interests ascertained and developing fair ways to resolve differences in interests—can also be applied in situations where there is no conflict, but where people seek to understand and accommodate each other's motivations. Interestingly, while one of us (Gabriele Bammer) has used principled negotiation in this way in large collaborative projects, we have been unable to find any documented examples of its use as a research integration tool. To assist the reader to understand how readily each method can be transposed to research integration, we provide a genealogy of the method and a commentary on its use in research integration in the description of each method.

While this is the first published compilation and analysis of dialogue methods for research integration, other sources cover some of the methods dealt with here and additional methods that we have excluded from this book, having judged that they are either not dialogue methods or are not useful for research integration. They apply quite different classificatory schemes. Examples include the following, and a fuller list is in Appendix 2:

- Start and Hovland (2004), *Tools for Policy Impact: A handbook for researchers.* Some 31 tools are covered in this source, classified into research tools for policy impact, context assessment, communication and policy influence.
- Carson and Gelber (2001), *Ideas for Community Consultation: A discussion on principles and procedures for making consultation work.* This source includes four of the methods we have covered, but its focus is community consultation.
- Keating (2002), *Facilitation Toolkit: A practical guide for working more effectively with people and groups.* This includes 20 tools. While the facilitation of dialogue is an important component of many of the methods we describe, our focus is not on facilitation as such, as is the case in Keating's publication.
- Urban Research Program, Griffith University (2006), *URP Toolbox.* This 'toolbox' contains 63 tools that can be used to improve the quality of stakeholder involvement in decision making, particularly regarding environmental sustainability. Again, it covers some of the dialogue methods discussed in this book.

Appendix Table 3.1 provides an extensive list of methods—some drawn from these publications—that we have used as a starting point for identifying dialogue methods for research integration.

Applying the dialogue methods in this book

Flexibility

As we pointed out in the section on classification, some of the methods are broadly applicable, while others are more narrowly targeted. We have suggested that the latter methods are used when an in-depth focus on a particular aspect of knowledge—such as interests or world views—is especially apposite. Experienced research integrators can also combine methods in helpful ways. For example, in the process of using a broad method, it could become evident that differing values or some other attribute are blocking the development of shared understanding, so that a method to specifically deal with this could be gainfully combined with the broad method. Thus, methods can be used in conjunction with others, either sequentially or nested. In the case we present on seeking agreement on the core operational strategy of a Cooperative Development Agency in the United States (see under strategic assumption surfacing and testing), it was recognised that reconciling two conflicting sets of assumptions regarding top-down versus bottom-up approaches was essential for moving forward. In this case, strategic assumption surfacing and testing was used to make clear the assumptions of the two main groups of stakeholders. This was nested within a soft systems methodology approach, which aimed to develop more general joint meaning and understanding. Combinations of the general techniques could also be useful. For example, a case study we describe integrating judgments for dealing with *Salmonella* infection started with the nominal group technique and followed it with a Delphi technique (the example can be found under Delphi technique), drawing on the different strengths of each method for particular aspects of the problem they were addressing.

Such flexibility in application of the techniques is critical for successfully using dialogue methods for research integration. It is the mark of a successful research integrator to be able to do this and such skill is built through training and experience. By endeavouring to provide a more systematic approach to dialogue methods for research integration, we are not seeking to undermine this vital flexibility in application. Instead, we aim to enhance it, by broadening appreciation of the range of available methods, as well as providing numerous examples illustrating how the methods have been applied.

Preparing to use a dialogue method

It is also worth noting that using many of the dialogue methods for research integration involves significant preparatory work. Further, some dialogue methods involve a series of meetings, interspersed with other activities. Some also require substantial action after the event to finish the integrative task. While our focus in the descriptions that follow is on the dialogue event itself, we also flag these other aspects.

Areas not covered in this book

The book does not provide some of the essential ingredients for successfully applying these dialogue methods, such as facilitation and other group management skills. For example, it does not consider important areas such as managing power differences between participants, managing intransigent participants or keeping to time limits. Our primary audience will already have many of these skills. For novices, this compilation is intended to be used in conjunction with training by experienced experts.

Furthermore, the book does not deal with critical areas such as the selection of participants or taking action based on the results of the dialogue; these are covered by other aspects of Integration and Implementation Sciences, particularly 'framing, scoping and boundary setting' and 'providing research support for decision making' (see Appendix 1).

How to read this book

This book has opened with an introductory and framing discussion and a clarification of what it covers and what is out of its scope. The next two chapters present the 14 dialogue methods, illustrating their role in research integration. The concluding chapters discuss differentiating between the methods—clarifying which methods are particularly useful for which integrative challenges—and the appendices place the dialogue methods into a broader context of Integration and Implementation Sciences.

Our descriptions of each of the dialogue methods are accompanied by one or more examples of their use in research integration. These examples are structured around six questions that we have found to be helpful in thinking systematically about research integration and documenting its application.

1. What was the integration aiming to achieve and who was intended to benefit?
2. What was being integrated?
3. Who did the integration?
4. How was the integration being undertaken?
5. What was the context for the integration?
6. What was the outcome of the integration?

As we demonstrate in the cases that follow, the questions can be used in any order, and can be combined. Further details on the use of this descriptive and analytic framework are provided in Appendix 1 and Bammer (2006a).

In Table 2.1, we provide an overview of how well the examples illustrate each particular method. First, we document the range of topic areas in which we have been able to find examples and where we had to resort to examples in areas outside environmental management, public health, security and technological innovation, or outside research integration. Second, we describe the participant groups each method is primarily useful for—that is, discipline and stakeholder experts, discipline experts only or stakeholders only—and which of these are illustrated by the case studies. Third, we describe whether the research role in the example is clearly integrative.

In Table 2.2, we describe some additional characteristics of each method:

a. the usual number of participants
b. the characteristics of the dialogue process
c. whether the locus of control lies with the participants or the organisers
d. how highly structured the method is
e. the extent to which preparatory or integrative work outside the dialogue is required
f. particular strengths
g. major limitations.

Table 2.1 How well the examples illustrate each particular method

Method	Cases					Participants			Research integrator role clear?
	Environment	Public health	Security	Technological innovation	Other	Disciplines and stakeholders	Disciplines only	Stakeholders only	
I. Dialogue methods for understanding a problem broadly: integrating judgments									
Citizens' jury	√	√				Yes			Yes, if the organiser
Consensus conference				√		Yes			Yes, if the organiser
Consensus development panel		√					Yes		Yes
Delphi technique	√	√	√	√		Yes			Yes
Future search conference		√				Yes			No
Most significant change technique	√							Yes	Yes
Nominal group technique	√	√				Yes			Yes
Open space technology	√	√				Yes			Yes
Scenario planning	√				Business		Yes?		Yes
Soft systems methodology			√			Yes			Yes
II. Dialogue methods for understanding particular aspects of a problem: integrating visions, world views, interests and values									
Appreciative inquiry		√			Organisational development	Yes			No
Strategic assumption surfacing and testing					Business and not research integration	Yes			Yes, if facilitator
Principled negotiation		√ (social work example)				Yes			No
Ethical matrix				√		Yes			Yes, if organiser/facilitator

Table 2.2 Additional characteristics of each method

I. Dialogue methods for understanding a problem broadly: integrating judgments

Method	Usual number of participants	Characteristics of dialogue process	Locus of control (participants or organisers)	Degree of structure in the method	Requirement for additional preparatory or integrative work outside the dialogue	Particular strengths	Major limitations
Citizens' jury	18–24 citizens, a microcosm of the public	Meet for 4–5 days, hear from expert witnesses, deliberate and present recommendations on final day	Participants	Highly structured	Significant preparation	Efficient Develops informed inputs to decision making	Only the views of citizens are elicited, not other stakeholders
Consensus conference	12–25 citizens, a representative sample of the public	Need for a preparatory weekend then 2–4 days	Participants	Highly structured	Significant preparation	Efficient Develops informed inputs to decision making	Only the views of citizens are elicited, not other stakeholders
Consensus development panel	About 15 panellists plus an open number of conference participants	Panel receives inputs from expert speakers over 1½ days, develops a draft consensus statement, discusses it with conference participants and releases it to the public	Organisers	Highly structured	Significant preparation Comments on draft consensus statement invited after the dialogue	Independent panellists synthesising a body of research evidence and producing their consensus position on it	Applicable only where a substantial body of scientific evidence has been published on a topic, and where the level of controversy is not so great as to preclude its synthesis and the panel producing a consensus statement
Delphi technique	Varies from just a few to hundreds	Operates by mail, email or internet. Participants respond to organisers' questions; responses are shared; usually three rounds	Organisers	Highly structured	Significant preparation	Taps knowledge and judgments of experts while avoiding the dominance by particular individuals that can occur in face-to-face dialogue	Resource intensive Organisers require significant management and integrative skills

Table 2.2 (continued)

Method	Number	Process	Determined by	Structure	Preparation	Benefits	Issues
Future search conference	Varies from about 60 to hundreds	Round-table plenary and small group discussions leading to the development of agreed visions and action plans	Organisers re process, participants re contents	Highly structured	Limited preparation but strategies needed for follow-up—that is, implementation of action plans	Can integrate disciplines and stakeholders. Focuses on post-conference action	Participants might not be able to find consensus re the nature of the problem and/or actions needed. Commitment to follow-up action could be strong at the conference but dissipate soon afterwards
Most significant change technique	Scores to hundreds	Change stories gathered in group discussion or in writing and reviewed at various levels in a hierarchical organisation	Organisers	Highly structured	Significant preparation, implementation management and follow-up	Informs senior managers. Gives voice to the less powerful stakeholders. Focuses on program outcomes and their drivers	Other techniques—dialogic and other—needed to gain a full understanding of the situation, outcomes, attribution and future action needed
Nominal group technique	Small groups of up to about 12	Face-to-face small group dialogue to generate, record, discuss and vote on ideas in such a manner as to minimise power differentials between participants	Organisers	Highly structured	Little required	Avoids the dominance by particular individuals that can occur in face-to-face dialogue	Requires a degree of shared understanding of the problem and willingness to listen and compromise
Open space technology	Any number, from a small group to thousands	Participants work together in small groups with like-minded people on topics they have identified as priorities	Participants	Fairly unstructured	Little required	Diversity among the participants. Encourages creativity and lateral thinking. Can produce action plans for implementation after the OST event	Requires clarity about the issue being addressed and willingness to listen and compromise. The unstructured nature of the OST events is problematic to some potential participants

Table 2.2 (continued)

Method	Usual number of participants	Characteristics of dialogue process	Locus of control (participants or organisers)	Degree of structure in the method	Requirement for additional preparatory or integrative work outside the dialogue	Particular strengths	Major limitations
Scenario planning	Variable, but not so many as to impede small group processes	Expert facilitators guide small group discussions, which produce the scenarios	Organisers	Highly structured	Detailed documentation of the scenarios is often undertaken outside of the dialogue	People with expert knowledge about a field address uncertainty	Selection of participants tends to shape the outcomes of the process Challenges exist in linking the scenarios developed to decision making and action planning
Soft systems methodology	Variable, but not so many as to impede small group processes	Participants engage in debate to understand others' world views and perceptions of a problem in its context, and then develop action plans to address it	Organisers	Can be highly structured (Mode 1) or more free-flowing and adaptive to circumstances (Mode 2)	Little required	Valuable in developing action plans to deal with complex social situations where the nature of the problem, its origins and what to do about it are unclear	The surfacing of participants' world views, and discussing their implications for understanding and addressing an issue, is challenging to some participants The method is not widely known and relatively few examples of its application and outcomes are documented

Table 2.2 (continued)

Method	Usual number of participants	Characteristics of dialogue process	Locus of control (participants or organisers)	Degree of structure in the method	Requirement for additional preparatory or integrative work outside the dialogue	Particular strengths	Major limitations
II. Dialogue methods for understanding particular aspects of a problem: integrating visions, world views, interests and values							
Appreciative inquiry	Varies from a small number to scores	Typically involves a work team engaged in small group dialogue to develop shared visions	Organisers	Structured	Preparation required to orient participants to the AI perspective	Can be a valuable organisational development tool that focuses on the future of the organisation	Participants need to be oriented towards ignoring current problems with the organisation's operations and, instead, focus on its future
Strategic assumption surfacing and testing	Varies from a small number to scores	A number of small groups with common assumptions, and dialectic debate plenary sessions	Organisers	Structured	Significant preparation	As a planning tool, it can reveal the diverse assumptions held by participants and find accommodations between them	Participants must be willing to expose, through dialogue, their underlying assumptions and have them challenged through dialectic debate
Principled negotiation	Generally small numbers (often two people) but can be two or more negotiating teams of any functional size	Willingness, on the part of all participants, to understand the interests that the others bring to the negotiating table, and to find a position acceptable to all	Participants	Moderately structured	Little preparation	Can eliminate conflict between the participants, leading to a mutually acceptable resolution	Participants sometimes find it difficult, or they are unwilling, to separate the people from the problem and to focus on participants' interests, not positions
Ethical matrix	Small groups	Round-table discussion to identify and reach consensus on the ethical implications of the issue being addressed and the relative importance of those implications	Participants	Highly structured	Can involve little or a lot of preparation, depending on the approach taken	Reveals the values that participants hold or ascribe to stakeholders not present, and weighs their relative importance using a structured framework	Participants must be willing and able to discuss value issues, and to empathise with stakeholders not at the table, identifying and analysing the values and ethical issues that underlie the topic

Key source documents for each method are provided as part of its description to assist readers wishing to further investigate particular methods, including developing skills in applying them. Literature citations provided within each section are detailed in the list of references that concludes this book.

Further comment on the examples presented in this book to illustrate different dialogue methods is also warranted, especially as the examples are intended to help readers think about how the methods can be applied. We present the best examples we could find and, while we could not search all the literature, we did attempt to cover a broad swathe of research publications (see Appendix 3). For some methods—for example, the Delphi technique—we were spoilt for choice. We found examples in each of our four areas of application and for various ways of combining discipline and stakeholder inputs, so that we could illustrate a range of ways of applying the method in research integration. More commonly, however, there are gaps in our illustrations. We usually could not find an example in each of the areas of environment, public health, security and technological innovation. More importantly, the examples of research integration that are demonstrated are often limited and, for some methods such as principled negotiation, non-existent.

We also note that most of the examples we have found concentrate on stakeholder input. Examples where different disciplinary or expert perspectives were brought together were less comon, and illustrations combining disciplinary and stakeholder inputs were rare. That is not to say that the participants in dialogue for research integration always have to conform to a particular stereotype. On the contrary, the point we are making here is that the illustrations we are able to provide cover only a limited array of possibilities in terms of bringing various perspectives together.

In our search for examples, wherever possible, we chose those where researchers were prominent: in organising the dialogue, as facilitators, as participants, as 'expert witnesses' and/or in documenting the dialogue. Because the role of researchers as integrators is not, however, yet well defined or established—for example, through a crosscutting discipline of Integration and Implementation Sciences—the tasks of the researchers in our examples are not always integrative or even clearly described.

Overall, we focus on description of dialogue methods, rather than analysis or evaluation. This reflects the fact that little analysis or evaluation of individual methods has been undertaken and published with respect to dialogue, let alone comparative analyses. Towards the end of this book, however, after we have presented each method, we take a first analytical step. We use a hypothetical problem based on concerns about amphetamine use in young people to illustrate an aspect of the problem each dialogue method is ideally suited to address. For the dialogue methods aimed at providing a broad understanding of a problem, we then tabulate which other methods can be used to address that aspect of the problem. Our aim is to help readers begin to differentiate between dialogue methods, allowing them to choose those most appropriate for a specific research integration task.

As we have outlined in the introduction, we see this book as charting new territory in linking dialogue methods to research integration. While this book is as comprehensive as we can make it based on published material, gaps and limitations remain, as we outline above. We believe, however, that we have demonstrated 'proof of concept', and that further attention to this area is likely to be worthwhile and productive. Considerable scope exists for further development of dialogue methods for research integration and for researchers as Integration and Implementation Sciences specialists. Our aim here is to lay the foundations for that development.

Chapter 3. Dialogue methods for understanding a problem broadly: integrating judgments

The majority of methods we identified were useful for the integration of judgments. Here, we define judgment as the 'ability to judge justly or wisely, especially in matters affecting action; good sense; discretion [and] the forming of an opinion, estimate, notion, or conclusion, as from circumstances presented to the mind' (Macquarie Dictionary 2005).

As we described earlier, Yankelovich (1999) went further than this, pointing out that, in making a judgment, people took into account the facts as they understood them, their personal goals and moral values and their sense of what was best for others as well as themselves.

In tackling real-world problems, it is common that research data alone are not sufficient to provide full understanding of the problem and a clear path for action. In addition, action often needs to be taken before all the research that can have a bearing can be conducted. Synthesising a range of informed judgments is then often the best way forward.

In research integration, the focus of the dialogue process is on a research question and the process aims to enable the formation of a combined judgment between the participants, with that judgment being informed by the best research evidence. Research-informed judgments can be achieved in various ways. One is to present research evidence to those whose judgments are being synthesised—through, for example, presentations, documentation or questions and answers. This is commonly done in dialogue methods that concentrate on integrating the (informed) judgments of lay people (citizens' jury and consensus conference). Another is to concentrate on research experts and to integrate their judgments (for example, in a Delphi technique). Still another is to involve research experts and lay people and to share the research evidence through discussion (for example, in open space technology).

The research integrator is most likely to take the lead in organising the dialogue and in bringing the results to the attention of decision makers. Tasks undertaken by research integrators can therefore include determining the topic for the dialogue and the particular dialogue method, the selection of participants, as well as what research evidence will be presented and how. They are likely to also be responsible for documenting the outcomes of the dialogue and ensuring that the process is evaluated, as well as deciding to whom the results should be presented and how that is best done.

Citizens' jury

Description

A citizens' jury is a dialogue method that was developed by the Jefferson Center in the United States. The centre has registered the term 'Citizens Jury' as a trademark in that country. This method is used by organisations wishing to receive and understand the views on complex issues of a well-informed, representative group of ordinary citizens.

The process involves providing the citizens with information from subject-matter experts, advocates and other stakeholders and then bringing together the range of judgments of the citizens into a single judgment.

The core approach (as used by the Jefferson Center) is as follows:

> In a Citizens Jury project, a randomly selected and demographically representative panel of citizens meets for four or five days to carefully examine an issue of public significance. The jury of citizens, usually consisting of 18 to 24 individuals, serves as a microcosm of the public. Jurors are paid a stipend for their time. They hear from a variety of expert witnesses and are able to deliberate together on the issue. On the final day of their moderated hearings, the members of the Citizens Jury present their recommendations to the public. (Jefferson Center 2004:3)

This source explains that the main characteristics of a citizens' jury are:

1. *representative*: selected by a recognised sampling method
2. *informed*: witnesses present to the jury a variety of facts, information and opinions on the matter under consideration, and are questioned by the jury
3. *impartial*: those organising the process select witnesses whose evidence is carefully balanced to ensure fair treatment to all sides of the issue
4. *deliberative*: the jury deliberates in a variety of formats and is given sufficient time to ensure that all of the jurors' opinions are considered.

As originally designed, the process operates in nine stages (Jefferson Center 2004), but as the cases below show, variations are possible.

1. *Establishment of an advisory committee* of four to 10 people with sound knowledge of the issue to be deliberated on. They advise the organisers about focusing the topic, the selection of witnesses and development of the agenda.
2. *A telephone survey* is conducted of a random sample of the public to obtain demographic and attitudinal information on the topic under consideration. Those polled who express interest in the topic are sent information about it, and about the citizens' jury process.

3. *Jury selection* occurs using techniques that aim to ensure that the jury is representative of the community from which it is drawn. Potential jurors are people who have been selected in the previous step and have agreed to have information on the topic and the jury process sent to them. They are categorised on the basis of demographic and attitudinal variables and jurors and alternatives are then selected to ensure representativeness.

4. *Witness selection* is the next step. The aim is to involve neutral resource people, as well as advocates and stakeholders. Care is taken to ensure balance in the witnesses' inputs.

5. *The charge* is determined—that is, the question or questions that the jury will consider and on which they will reach a judgment. Care needs to be taken to ensure that the scope of the charge is neither too narrow nor too broad.

6. *The hearings* are then conducted and moderated by professional facilitators. The staff prepare the venue, the order of witnesses, and so on. The hearings typically run all day for five consecutive days. Ample time is provided for jurors to discuss the issues among themselves (in small and large groups) as well as with the witnesses. The presentations end on the afternoon of the fourth day. On the morning of the fifth day, the jurors have their final discussions and prepare an answer to the charge—that is, determine their judgment. They also review an initial report from the process.

7. *Recommendations* are issued by the jury members, along with their findings, at a public forum on the final afternoon.

8. *Evaluation*, which involves all the jurors, is undertaken, with an important evaluation question assessing whether or not they feel that the process has been biased in any way. They are also invited to write a personal statement about the process. The evaluations and personal statements are included in the final report.

9. *Public outreach* occurs throughout the life of the project, sometimes entailing a web site with transcripts of evidence and media liaison activities to promote public interest in the process and awareness of its conclusions.

The Jefferson Center met difficulties that reflected aspects of the US political system and closed in 2002. (The Internal Revenue Service revoked the centre's tax-deductible status and it was threatened with legal action owing to its work in evaluating election candidates' policies.) Nonetheless, citizens' juries continue to be conducted in the United States and other countries, notably Germany, the United Kingdom, Denmark, Spain and Australia.

Examples of its use in research integration

1. The environment: deciding the future of a wetland

What was the context for the integration?

The Fens is a large, low-lying area near Ely, Cambridgeshire, in the United Kingdom. It was formerly a wetland (hence its name), but it was drained and turned into arable farmland, although there were moves to return at least some of the area to the original wetland state.

What was the integration aiming to achieve and who was intended to benefit?

A citizens' jury was conducted by researchers to explore the question 'What priority, if any, should be given to the creation of wetlands in the Fens?' (Aldred and Jacobs 2000). The organisers used the citizens' jury approach to dialogue because '[t]he premise behind this method is that, given enough time and information, ordinary people can make decisions about complex policy issues' (Aldred and Jacobs 2000:218).

What was being integrated?

The process aimed to integrate the judgments of 16 lay people (jurors), 'acting in their capacity as citizens concerned with the public good rather than [as] consumers concerned with private interests' (Aldred and Jacobs 2000:217) once they became well informed about the issues as presented by witnesses. The report on the jury process did not identify the characteristics of the witnesses and the topics on which they made presentations, stating simply that they were 'experts on different aspects of the question being considered' (Aldred and Jacobs 2000:220).

How was the integration undertaken and who did the integration?

This citizens' jury was freestanding in the sense that it was not commissioned by a body that had the power to implement the jury's findings, as would occur in most traditional citizens' juries. (Such a freestanding nature is likely to be common in research integration.) It was established, however, with the support of an advisory group made up of representatives from organisations with varying interests in the development of the Fens—that is, a range of stakeholders. Although the members of the advisory group did not commit themselves to be bound by the jury's conclusions, they 'agreed to receive the report and take its recommendations seriously' (p. 219). The question deliberated on by the jury was developed jointly by the researchers and the advisory group, as were the options that the jurors were asked to discuss. Some of the witnesses who gave evidence to the jury were representatives of the organisations on the advisory group.

Sixteen jurors were selected from among members of the public living in the local area. As the researchers who conducted the project pointed out, they had first to decide whether to choose a representative sample based on statistically random selection (impossible if there were to be just 16 jurors, they concluded) or whether to obtain a jury that was representative of the various local interest groups. They adopted the second of these options. In doing so, however, they explicitly excluded from selection people with strongly held views on the specifics of the topic. Instead, they selected a jury that included all relevant perspectives on the issues. This was done by having a market research firm interview local residents, telling them that the jury would

> discuss issues that are important matters of concern to people living locally. These issues will include farming, job creation, tourism, wildlife and the environment in Cambridgeshire, Norfolk and the Fens. The Jury is supported by Cambridgeshire and Norfolk County Councils and a group of other public bodies…[They were advised that] [t]o be a member of a CJ, you need no special training or experience. You are being asked to contribute your views and opinions as an ordinary member of the local community. (Aldred and Jacobs 2000:219)

Jurors were each paid £200 plus travelling expenses for four days, and were asked to sign a contract committing themselves to attend each day and to participate fully.

The agenda, prepared by the researchers and the advisory group, was presented to the jurors on the first day. It set out the overall question to be tried ('What priority, if any, should be given to the creation of wetlands in the Fens?') and covered four options for developments in the Fens area: turning the farmland into a nature reserve; establishing a wetland as a Fen centre; incremental development; or not taking any deliberate action at all. The first three options were proposals being advocated by various local organisations, and the fourth gave jurors the option of rejecting the idea of a wetland area in the Fens.

Following the usual procedure, a number of expert witnesses (selected mostly on the advice of the advisory group) made presentations to the jury and were questioned closely by its members. The 16 jurors regularly split into three groups (with varying membership) over the four days to consider particular topics in detail. This took the integrative work to a greater depth. The deliberations were captured by means of summaries prepared by the jurors on flip charts during small group work and plenary sessions, written records kept by the facilitators, tape recordings (everything was recorded) and evaluation questionnaires completed by the jurors at home after the jury concluded its work.

What was the outcome of the integration?

The jury's conclusions and recommendations were unanimously in support of option one (establishing a nature reserve in the Fens) and option three (incremental development). Careful attention was given to, and recommendations made about, a range of related issues, some of which were raised by jurors themselves—that is, matters that were not included in the pre-prepared agenda.

The jury process apparently worked well:

> The design and execution of the Ely Jury was in general considered very successful. The atmosphere in the meeting room over the 4 days was appropriately relaxed, with much laughter and all jurors appearing at ease to speak. All jurors were given an evaluation form on the final day; the comments on the forms returned (14 out of 16) were generally very positive. Most of the significant criticisms raised had already been recognised and acknowledged by the researchers. (Aldred and Jacobs 2000:222)

The full report on the Ely Citizens' Jury is available online (<http://alba.jrc.it/valse/pdf files/rap.fin.chapter6-a.pdf>). No information was provided on how successful the process was in informing decision makers or what action, if any, was taken based on the citizens' jury findings and recommendations.

2. Public health: planning local health and welfare services

What was the context for the integration?

South West Burnley is a town in Britain characterised by high levels of social exclusion, social deprivation and complex community needs regarding health, housing, employment, education, and so on. Over the years, a number of needs assessments had been conducted, mostly using consultation with the residents, but nothing came from them. In 1999, the then Burnley Primary Care Group commissioned a citizens' jury. Oversight of the process was the responsibility of a community-based Health and Social Care Group, which had close links to the Primary Care Group, as well as action researchers from Lancaster University.

At the time, Primary Care Groups were part of the UK National Health Service's Area Authorities (they have now been replaced by Primary Care Trusts). Each Primary Care Group organised and funded primary healthcare services within a particular region, either through other National Health Service organisations or through other bodies including private medical practitioners and not-for-profit organisations. This meant that the Primary Care Group had considerable capacity to implement the recommendations of the citizens' jury.

What was the integration aiming to achieve and who was intended to benefit?

The citizens' jury approach was selected for three reasons:

- to build up a body of evidence through the testimonials of jurors, witnesses and other members of the community
- to resolve a specific longstanding problem regarding local service provision

- to open up debate about what people living in that community would prioritise for change (Kashefi and Mort 2004).

The authors of the report on the citizens' jury added that, '[a]s action researchers, we were keen to see whether the jury process could hold service providers and policymakers accountable to the community, and to explore whether the process could be used as a tool for health activism' (Kashefi and Mort 2004:292).

Importantly, the key planning and service agencies agreed that this would be the final consultation and that action would flow from its findings. Deferring action on the community's pressing concerns was no longer a viable option.

The steering group, drawn from the Health and Social Care Group, set the question to be tried by the jury: 'What would improve the health and well-being of the residents of South West Burnley?'

What was being integrated?

There is little in the report about the information provided to the jurors as part of their deliberations.

How was the integration undertaken and who did the integration?

An innovative approach to juror selection was used. The steering group analysed census data on the demographics of the area and determined the characteristics jurors should represent. A professional recruiter, who had worked in the area previously, spent some weeks informally discussing the proposal with community members with the goal of finding potential jurors who matched the profiles set by the steering committee. A diverse group of 12 local residents, aged from seventeen to seventy years, was eventually recruited as jurors. They had two evening preparatory sessions at which they got to know each other and the other people involved in the process, including the researchers. A local community worker and four people who had served on a previous citizens' jury in the area designed a training session on the practical aspects of being a juror. The jurors were paid for their participation.

The deliberative phase of the jury was conducted over five days. A wide range of witnesses gave evidence, including health workers, community development workers, community activists, general practitioners and a social worker. Most were local residents. There were two closed sessions in which the jurors focused on their own ideas and experiences. Four pieces of research were commissioned before the deliberative phase to bring local community views to the jurors and the resulting research reports were presented as evidence by various witnesses. Jurors engaged in dialogue in pairs, small groups and separate female/male groups. The process was supported by two facilitators, a chairperson and a 'jury investigator'. The last person's role was to research and find answers to questions that arose in the course of the trial, and to rapidly (that is, within the trial time) report back to the jury.

What is the outcome of the integration?

On day five, the key recommendations were collated and pairs of jurors presented them using simple visual representations. These presentations were made to the sponsors of the jury and the local Member of Parliament. The researchers prepared a written report in consultation with the jurors (see <http://www.lancs.ac.uk/fass/ihr/publications/elhamkashefi/burnley citizens' jury report.pdf>). The jury's report covered the strengths and weaknesses of the community and included some 80 recommendations—some rather general and others very specific.

The Health and Social Care Group established a system to manage the dissemination of the report to the local community and other stakeholders, and to manage the implementation of its recommendations. Some of the jurors were actively involved in this for an extended period. As the authors of the paper describing this citizens' jury advise:

> Three months after publication [of the jury's report], all health and social care agencies mentioned in the report attended a public meeting to respond formally to the recommendations. Hopes were high for this event, that concrete action would be reported and plans outlined to work on the wider concerns raised by the jury...every agency sent a formal, written response to the Health and Social Care Group outlining their proposed course of action on the recommendations. (Kashefi and Mort 2004:298)

Funding bids were submitted and a multifunction community health centre was established and staffed. This was a centre for continuing community participation and service delivery for overcoming many of the problems faced by Burnley's citizens. The researchers involved in the case describe the importance of integration, as follows:

> [By integration] we mean that the jury process must be embedded within the community where it happens. The subject under discussion must be of relevance to community groups and organizations, as well as individuals, and these should participate from inception to realization; knowledge from previous consultations must be integrated using local expertise, e.g. as witnesses and advisors; recommendations must be implemented using links with existing local networks in order for these to be credible and workable. Without integration, juries and other consultative activities will remain isolated and irrelevant. (Kashefi and Mort 2004:299–300)

Commentary

For our cases, we could find examples of research integration in only two of the four areas of interest (the environment and public health). Even so, it is noteworthy that the role of the researchers was not particularly clear and had to be inferred from the available documents. Even more importantly, in neither case was the evidence that was being integrated into the judgments described.

Use of the integration framework shows that, in terms of research integration, the available descriptions are very uneven. In both instances, the selection of the jurors is presented in considerable detail, while, as mentioned above, the information they were given is almost completely ignored.

This method was not developed for research integration; rather, it was aimed to be part of a broad democratic process in that it elicited and channelled information from concerned citizens to policy-makers, alerting them to the judgments of *informed* citizens on complex policy issues. Nevertheless, as the examples show, it can be used for research integration.

The method is valuable for integrating the informed judgments of citizens but not the judgments of other stakeholders, including those having to make decisions on the topic under consideration, and those potentially affected by those decisions. This means that it tackles only part of the research integration task. Other, complementary methods are needed to integrate the views of other stakeholders.

Origins and genealogy

The origins of this dialogue technique are in deliberative democracy. It is closely linked to the consensus conference (see below) and citizens' panels.

The citizens' jury method was developed at the Jefferson Center, Minneapolis, Minnesota, United States, which was founded in 1974 to undertake research and development on new democratic processes. The first citizens' jury was held the same year to pilot and explore the method; it addressed the topic of a national health plan for the United States. The centre closed in 2002 but still maintains a website <http://www.jefferson-center.org/> and conducted 32 citizens' juries before closing.

Further reading on citizens' juries

Carson, L., Sargant, C. and Blackadder, J. 2004, *Consult Your Community: A guide to running a youth jury*, NSW Premier's Department, Sydney, New South Wales.

Coote, A. and Lenaghan, J. 1997, *Citizens' Juries: Theory into practice*, Institute for Public Policy Research, London.

Crosby, N. and Nethercut, D. 2005, 'Citizens juries: creating a trustworthy voice of the people', in J. Gastil and P. Levine (eds), *The Deliberative Democracy Handbook: Strategies for effective civic engagement in the twenty-first century*, Jossey-Bass, San Francisco, pp. 111–19.

Jefferson Center 2004, *Citizens Jury Handbook*, Revised and updated edition, The Jefferson Center, <http://www.jefferson-center.org/>

Wakeford, T. 2002, 'Citizens juries: a radical alternative for social research', *Social Research Update*, no. 37, <http://sru.soc.surrey.ac.uk/SRU37.html>

Consensus conference

Description

The consensus conference is a highly structured event designed to involve non-expert, non-partisan citizens in deliberating on important (and typically complex) social, technological, planning and/or policy issues, and through doing so to integrate judgments. Their objective is 'to bridge the gap between the general public, experts and politicians, who only rarely have an opportunity to meet' (Grundahl 1995:31).

The chief characteristics of the consensus conference have been described as follows:

- the profile of participants is usually structured to provide a representative sample of the whole citizen group being consulted (by age, place of residence, gender, etc.)
- It involves relatively small numbers of participants (usually 12–25)
- It requires an independent and skilled facilitator
- Participants are provided with written evidence before they meet
- Participants decide who to call in as 'expert' witnesses, which allows the infusion of higher levels of knowledge and experience into the process
- It is interactive, participants meet for preparatory weekends and then a deliberative meeting of 2–4 days
- Recommendations are published in a formal report
- Either the recommendations are implemented, or sufficient grounds must be provided publicly to explain why they will not be implemented (Carson and Gelber 2001).

The consensus conference method was initially developed in Denmark, and the Danish approach has been applied widely. Some modifications have, however, been made in other settings. The steps involved have been summarised by Hendriks (2005:83):

> The Danish model is based upon a two-stage procedure that engages ten to twenty-five citizens in eight days of deliberation over a period of approximately three months. In the first stage, the citizens meet for two preparatory weekends to learn about the topic, the process, and the group. During these weekends, the panel [i.e. the citizens] also develops a series of questions for the conference to address and selects the conference presenters from a list of possible experts and interest group representatives.

> In the second stage of the process, the actual four-day conference takes place. On the first two days, various presenters appear before a plenary forum to respond to the questions set for the conference. Throughout

this period, the citizens' panel retreats into non-public sessions to formulate further questions for the presenters and to clarify any misunderstandings or points of contention. On the last two days, the citizens work together to write a report outlining their key recommendations, which they then present to relevant decision makers before a public audience. In some cases, the presenters have the right to reply, after which the citizens are free to reformulate their report.

In the Danish design, the people selected to be on the citizens' panel do not have expert knowledge of the topic and do not hold any strong views on it.

Consensus conferences tend to be run by government agencies or professional organisations. For example, the Danish Board of Technology has run many, with its conference reports being submitted to members of parliament. Australian examples include the Consensus Conference on Gene Technology, sponsored jointly by 28 government, research and industry bodies, and the annual Consensus Conference on Guidelines for the Use of Antiretroviral Agents in HIV-1 Infected Adults and Adolescents run by the Australasian Society for HIV Medicine. While these share some of the processes used in the US National Institutes of Health's Consensus Development Panel method (discussed below), the approaches used are more flexible, not following the highly structured format of the National Institutes of Health's approach.

Suitable topics for consensus conferences are matters of current interest that are: reasonably delimited and not too abstract, contain conflicts between the positions of various advocates, need clarification of objectives and the attitudes of proponents and opponents, require the contribution of experts to clarify the science and other issues underlying the topic, and for which the necessary knowledge and expertise are available (Grundahl 1995).

As with citizens' juries, in research integration, the consensus conference addresses a research question and is provided with a range of research evidence.

Example of its use in research integration

Technology: implications for the democracy of new telecommunication technology

What was the integration aiming to achieve and who was intended to benefit?

Technology assessment has traditionally been undertaken by subject-matter experts, but, in April 1997, a consensus conference was held in Boston to involve lay people. Its title was Citizens' Panel on Telecommunications and the Future of Democracy. This was the first consensus conference to be held in the United States and it was designed and implemented in a manner to assist lay people to learn about new technology and to develop an informed judgment about it. Its general goal was 'to improve decision making about science and technology by expanding access and perspectives beyond the traditional elite, increase the public understanding of science and technology through informed public debate, and enhance democracy by fostering civic engagement' (Guston 1999:452).

What was being integrated?

The staff of the project selected 16 experts to make presentations to the citizens' panel: four were academics, three were from government, two were from the not-for-profit sector and seven were from the corporate sector. Six of them were also on the project's steering committee. The report on the conference does not provide details of the topics that they covered.

How was the integration undertaken and who did the integration?

The consensus conference was sponsored by a number of prestigious organisations, including universities, *Technological Review* magazine, the Loka Institute (a non-governmental organisation) and the US National Science Foundation. It had a four-member directorate drawn from the main sponsors and the directorate established a 12-person steering committee comprising academics, activists and representatives of expert practice, sponsors and targeted groups.

To identify the 15 consensus conference panellists, some 1000 people in the Boston area were randomly selected from the telephone directory and phoned. The 127 people who expressed interest were mailed background information and a questionnaire, and were selected on the basis of their responses. Additional members were targeted to attain diversity in race, age, educational level and computer familiarity. The panellists had to commit to seven days of involvement in the process, spread over three weeks: two preparatory weekend sessions and the three-day public conference. The panellists selected the subtopics for discussion and 'described the kinds of information and expertise they wanted the steering committee and project staff to gather' (Guston 1999:456) and to have presented at the public conference.

The three-day public conference included presentations by 16 selected experts and questions from the 15 panellists. On the final day, the panellists gave a media conference during which they promulgated their four-page consensus statement. Media coverage of the conference, particularly by local media outlets, was extensive.

What was the context for the integration?

The key driver for the consensus conference was that the US Federal Communications Commission was developing policies in the area of emerging telecommunication technologies, particularly with respect to the US Government's goal of access to the internet for all Americans. This prompted initial action to establish the consensus conference from the Education for Public Inquiry and International Citizenship program at Tufts University in Boston.

What was the outcome of the integration?

The consensus conference was evaluated on four criteria: direct, instrumental impact; impact on general thinking in policy arenas; impact on the training of knowledgeable personnel; and the interaction of the analysis presented by the experts with the lay knowledge of the panellists (Guston 1999).

The integrative aspect of the process was seen particularly in the fourth of these:

> The interaction of analysis with lay knowledge is perhaps the most interesting category of potential impacts of a consensus conference. The presence of lay citizens on the panel means that the interaction with lay knowledge pertains to both participants—the panel members—and non-participants alike. (Guston 1999:469)

Panellists reported that they had learnt a lot about the substantive issues covered, were sensitised to the issues and learnt about civic involvement in policy issues. The experts reported how valuable they had found the interaction with the lay panellists, in contrast with their usual, narrower circle of colleagues. The evaluation concluded that the integration might have been even stronger if the experts and panellists had been given more opportunities for informal dialogue at the conference, perhaps during mealtimes and other breaks, to supplement the formal presentations and question and answer sessions.

In contrast, no real (direct, instrumental) impact was observed. The authors concluded that:

> The single greatest area of consensus among the [evaluation] respondents was that the Citizens' Panel on Telecommunications and the Future of Democracy had no actual impact. No respondent, not even those governmental members of the steering committee or expert cohort, identified any actual impact. The principal reason for this finding is that having an actual impact was not a primary goal of the citizens' panel...Many respondents, however, felt that an actual impact would have been desirable. Some lamented its absence. (Guston 1999:462)

Commentary

The consensus conference method has a role in integration when the goal is to have citizens develop and communicate, to decision makers, their informed judgments on a topic of public policy interest. The presence of the decision makers who will receive their findings is a distinguishing characteristic of the method. The peripheral involvement of researchers (for example, as organisers, presenters or evaluators, rather than the people making the integrative judgments) means, however, that it is not as closely tied to research as some other methods.

The example provided adheres closely to the standard approach for this method, especially with respect to the selection of panellists from the community at large and the extensive dissemination of its findings via the mass media.

The method can be contrasted with the similarly named consensus development panel method (described below), used primarily by the US National Institutes of Health (NIH). In that approach, the panellists are selected for their expertise, in their own scientific disciplines, rather than attempting to be representative of the community.

Origins and genealogy

The Danish Board of Technology developed the consensus conference model in the late 1980s. Its starting point was the US NIH's Consensus Development Panel model described below, but they altered it to eliminate from the panel people with expertise in the area being investigated, so as to involve only lay citizens representative of the general community.

The first Australian consensus conference was conducted at the Australian Museum in March 1999 on the topic of Gene Technology in the Food Chain (details: <http://www.amonline.net.au/consensus/> and <http://www.abc.net.au/science/slab/consconf/>).

The consensus conference approach to dialogue is closely related to citizens' juries (see above). Where they differ, however, is that the former generally takes place over a far longer period (months not days), the agenda and questions to be put to the experts are developed by the panellists and they also determine which experts they wish to have present to them (Carson and Gelber 2001).

Further reading on the consensus conference

Hendriks, C. 2005, 'Consensus conferences and planning cells', in J. Gastil and P. Levine (eds), *The Deliberative Democracy Handbook: Strategies for effective civic engagement in the twenty-first century*, Jossey-Bass, San Francisco, pp. 80–110.

Joss, S. and Durant, J. (eds) 1995, *Public Participation in Science: The role of consensus conferences in Europe*, Science Museum with the support of the European Commission Directorate General XII, London.

Consensus development panel

Description

The consensus development panel and the associated consensus development conferences are a dialogue method for research integration developed and implemented under the Consensus Development Program of the US NIH.[1] Their purpose is to provide guidance in areas of medical and broader health practice, particularly in areas in which controversy exists and a body of scientific evidence is available that can be scoped, explored, assessed and synthesised to produce a consensus statement on the issue.

The NIH lists six principles that govern the conduct of a consensus development program conference and the operation of the consensus development panel:

1. A broad-based, non-DHHS [US Department of Health and Human Services], non-advocacy, independent panel is assembled to give balanced, objective, and knowledgeable attention to the topic. Panel members are carefully screened to exclude anyone with scientific or financial conflicts of interest...

2. Invited experts present data to the panel in public sessions, followed by inquiry and discussion. The panel then meets in executive session to prepare the statement.

3. Four to five predetermined questions define the scope and direction of the conference. These questions are widely circulated and are known to all conference participants. The principal job of the panel is to develop responses to them.

4. A systematic literature review is prepared for use by the panel in addressing the questions. The review is prepared by the Agency for Healthcare Research and Quality (<http://www.ahrq.gov/>) through one of several Evidence-based Practice Centers [of the NIH].

5. Near the end of the conference a draft conference statement is prepared by the panel in executive session and is then presented in plenary session. Following public discussion the panel may modify the statement as they deem appropriate and the resulting statement is posted on the website (<http://consensus.nih.gov/>) as DRAFT and is usually finalized in 4–8 weeks postconference.

6. The consensus statement is widely disseminated to achieve maximum impact on health care practice and medical research. (NIH 2008)

Two closely related types of conferences are conducted, each using the same approach but dealing with slightly different subject matter and concluding with slightly different consensus statements: the consensus development conference and the state-of-the-science conference. The difference between them is that the former covers areas of science and health practice for which a strong evidence base exists from randomised controlled trials and high-quality observational studies. The latter deals with areas where the evidence is weaker. The purpose of state-of-the-science conferences is not to answer specific research questions or resolve controversies (as occurs with the consensus development conferences) but to summarise the evidence and recommend future directions for research. Otherwise, the two types of conferences and panel processes are very similar.

From 2003 to 2007, the Consensus Development Program conducted 10 conferences, of which two were consensus development conferences and eight were state-of-the-science conferences. The conferences held in 2006 and 2007 illustrated the breadth of the topics covered: 'Prevention of Fecal and Urinary Incontinence in Adults', 'Tobacco Use: Prevention, cessation and control', 'Multivitamin/Mineral Supplements and Chronic Disease Prevention' and 'Cesarean Delivery on Maternal Request' (a full list is available online at <http://consensus.nih.gov/>).

The core outcome of the panel's deliberations is a consensus statement on the topic that they have addressed. This is disseminated widely to the public (for example, through media conferences and web-casts of conference proceedings) and to scientific and practitioner audiences (for example, through the NIH web site, mailings of the statement and articles in refereed professional journals). The statements are characterised as follows:

> A consensus statement is based on publicly available data and information. It is not intended as a legal document, practice guideline, or primary source of detailed technical information. Rather, the statement reflects the views of a panel of thoughtful people who understand the issue before them and who carefully examine and discuss the scientific data available on the issue. The creative work of the panel is to synthesize this information, along with sometimes conflicting interpretations of the data, into clear and accurate answers to the questions posed to the panel. The statement may reflect uncertainties, options, or minority viewpoints. Following the conference, the consensus statement receives wide circulation through both lay and medical media. Conference proceedings are webcast live (<http://consensus.nih.gov/>) and archived for later viewing. (NIH 2008)

The NIH's conference organisers select the panel chairperson and that person, in conjunction with the organisers, selects the other panellists. A planning committee is appointed to manage the process, including drafting the conference questions and program.

Integration is undertaken by the panellists and occurs through their work in synthesising and judging the data and research evidence that are presented in the comprehensive literature review prepared for them before the panel is convened, and which is presented in conference sessions by outside experts. These inputs, and the linked co-sponsorship of the conferences, frequently come from a number of disciplines and practitioner perspectives. An example is the 2004 state-of-the-science conference on 'Preventing Violence and Related Health-Risking Social Behaviors in Adolescents', which was co-sponsored by agencies as diverse as the US Agency for Healthcare Research and Quality, the Centers for Disease Control and Prevention, the National Institute on Alcohol Abuse and Alcoholism, the National Institute of Child Health and Human Development, the National Institute on Drug Abuse, the National Institute of Nursing Research, the National Library of Medicine, the Office of Behavioral and Social Sciences Research, the Substance Abuse and Mental Health Services Administration and the US Departments of Education and Justice.

Example of its use in research integration

Public health: developing a state-of-the-science statement on the prevention, cessation and control of tobacco use

What was the context for the integration?

In many societies, tobacco smoking is the leading preventable cause of premature death. Tobacco contains a highly addictive, though otherwise reasonably safe, chemical, nicotine. Many different strategies aiming to reduce the adverse health and societal impacts of tobacco smoking exist, including health education, controls on the physical availability of tobacco products, taxation and promoting less dangerous ways of ingesting nicotine. Uncertainties exist, however, about which strategies, or mix of strategies, are most likely to be effective among which population groups.

What was the integration aiming to achieve and who was intended to benefit?

A US National Institutes of Health State-of-the-Science Conference on the Prevention, Cessation and Control of Tobacco Use was conducted in 2006. As an integrative method, it aimed to synthesise research evidence and integrate the judgments of experts, to produce new knowledge or confirm existing knowledge in the light of new evidence, which would provide better understanding of the impediments to reducing smoking and tobacco-related health and social problems, and how to overcome these impediments.

As is standard with the NIH consensus conference and state-of-the-science conference methods, a small number of questions was identified for the panel to address:

- What are the effective population- and community-based interventions to prevent tobacco use in adolescents and young adults, including among diverse populations?

- What are the effective strategies for increasing consumer demand for and use of proven, individually oriented cessation treatments, including among diverse populations?
- What are the effective strategies for increasing the implementation of proven, population-level, tobacco-use cessation strategies, particularly by health care systems and communities?
- What is the effect of smokeless tobacco product marketing and use on population harm from tobacco use?
- What is the effectiveness of prevention and of cessation interventions in populations with co-occurring morbidities and risk behaviors?
- What research is needed to make the most progress and greatest public health gains nationally and internationally? (National Institutes of Health State-of-the-Science Panel 2006:839)

The intended beneficiaries included smokers and their families, and society at large, which was adversely impacted on by the monetary and social costs of tobacco use.

What was being integrated?

The panel's task was to judge and synthesise the existing data and research evidence to produce a clear statement of what was currently known in the first five of the areas listed above, and to identify which areas of new research were likely to have the greatest impact (question six). The background paper prepared for the panel by the US Agency for Healthcare Research and Quality (Ranney et al. 2006) was a systematic review that began with 1288 sources, demonstrating the breadth and depth of the research effort on tobacco to date. Just 102 of the initial sources met the review's inclusion criteria, particularly those addressing research quality.

How was the integration undertaken and who did the integration?

The integration was undertaken in two phases. The first was the development of a 421-page *Evidence Report* or systematic review of the scientific literature, undertaken by the US Agency for Healthcare Research and Quality. After confirming the findings of earlier reviews about intervention effectiveness, it concluded:

> The evidence base has notable gaps and numerous study deficiencies. We found little information to address some of the issues that previous authoritative reviews had not covered, some information to substantiate earlier conclusions and recommendations from those reviews, and no evidence that would overturn any previous recommendations. (Ranney et al. 2006:v)

The second stage was the work of the panel itself. The NIH convened a 27-person steering committee for the conference, chaired by the chief of its Tobacco Control

Research Branch. While the majority of the steering committee members were on the staff of the NIH, others came from outside institutions, predominantly universities.

The conference was conducted over three days, 12–14 June 2006, in a conference centre at the NIH in Bethesda, Maryland, United States. Media publicity preceded the event: the public was invited to attend and arrangements were made for a live web-cast of the proceedings. Mark Clanton, Deputy Director of the National Cancer Institute, opened the conference and the charge to the panel was delivered by Barnett S. Kramer, Director of the Office of Medical Applications of Research in the Office of the Director of the NIH. The panel and conference were chaired by David F. Ransohoff, Professor of Medicine at the University of North Carolina at Chapel Hill. The panel had 14 members drawn from the disciplines and fields of cancer prevention, nursing, health research, special population groups and various medical specialties. All were independent of the NIH, were not directly involved in tobacco research or closely allied fields and had not adopted an advocacy role relating to tobacco before the conference.

During its first day-and-a-half, the conference was addressed by 18 speakers from various disciplines. Opportunity for discussion followed each group of presentations. On the afternoon of the second day, the panel met in a closed executive session to begin its integrative work, synthesising and judging what they had read and heard and producing a draft state-of-the-science statement. Their task was to integrate the research evidence coming to them from the various disciplines and perspectives of conference participants and to integrate their individual judgments of this evidence to produce a whole-of-panel consensus statement.

On the morning of the third and final day, the panel presented to the conference its draft statement, inviting comments from the floor as it went through it section by section. The panel then met in a second closed executive session to review the public comments it had received. On the afternoon of the third day, a media statement was issued and a media conference convened to begin the dissemination of the panel's findings. The draft statement was immediately posted on the Consensus Development Program's web site and the final statement, after further deliberation by the panel, was released some weeks later (these documents are available online at <http://consensus.nih.gov/2006/2006TobaccoSOS029 main.htm>). As is standard practice for these conferences, the statement was subsequently published in the medical journal *Annals of Internal Medicine* (National Institutes of Health State-of-the-Science Panel 2006).

What was the outcome of the integration?

The huge amount of data and research evidence about interventions for the prevention, cessation and control of tobacco use was synthesised, through the conference process, enabling the panel to issue a state-of-the-science statement on the topic. To achieve this product, the panel members had to gain a sound understanding of the evidence, synthesise the multiple types of evidence addressing each individual component of their remit and weigh the conflicting and qualitatively different types of evidence. Furthermore, although the panel was relatively large, with 14 members, its resulting statement represented their consensus position; no dissenting reports were included.

The final statement was brief—19 pages—and was structured around the six questions reproduced above. For each, it stated why the question was important and summarised 'what we know' and 'what we need to know'. The statement concluded:

> Tobacco use remains a very serious public health problem. Coordinated national strategies for tobacco prevention, cessation, and control are essential if the United States is to achieve the Healthy People 2010 goals. Most adult smokers want to quit, and effective interventions exist. However, only a small proportion of tobacco users try treatment. This gap represents a major national quality-of-care problem. Many cities and states have implemented effective policies to reduce tobacco use; public health and government leaders should learn from these experiences. Because smokeless tobacco use may increase in the United States, it will be increasingly important to understand net population harms related to use of smokeless tobacco. Prevention, especially among youth, and cessation are the cornerstones of strategies to reduce tobacco use. Tobacco use is a critical and chronic problem that requires close attention from health care providers, health care organizations, and research support organizations. (NIH 2006:2)

Commentary

The consensus development panel process developed and used regularly by the US NIH is a highly structured approach for integrating scientific research evidence emanating from different disciplines. At the core of the process is judging the evidence and reaching a consensus position that is then communicated to the health professions and the public in non-technical language.

It shares some features with citizens' juries and the consensus conference, particularly in that experts and others present evidence to a panel, the members of which then synthesise and judge that information to reach a decision on the topic being explored. Where the NIH consensus development panel process differs, however, is that the panellists are not intended to be representative of the citizens of a community. Instead, they are selected on the basis of their advanced expertise in some area of health, as well as their independence from the sponsoring bodies and advocacy groups. They are not experts in the subject matter of the particular conference and have no financial or career advancement investment in the topic covered. As the Consensus Development Program explains, their role is more like that of a judge than a subject-matter expert (<http://consensus.nih.gov/FAQs.htm>).

The example provided above shows close adherence to the standard approach. This reflects the management of the process, from inception to conclusion, by the staff of the NIH's Consensus Development Program.

As an integrative method, it is particularly valuable when a body of research evidence is available addressing a tightly defined topic, and where experts in the area feel that the evidence needs to be—and can be—drawn together and weighed by professionals to develop a shared understanding of what the evidence reveals and what future research can be undertaken to fill gaps in knowledge. The independence of the panellists and the significant efforts made to draw to the public's attention the conference process and the panel's findings are also significant features.

Origins and genealogy

The Office of Medical Applications of Research at the US NIH conducts the Consensus Development Program. Conferences have been conducted under the program since 1977, with at least one conference held each year since then. We are not aware of any documentation explaining how the program originated.

Further reading on consensus development panel

The NIH's Consensus Development Program's web site (<http://consensus.nih.gov/>) provides detailed information on the goals and implementation processes used by the consensus development panel. It also provides full documentation of completed conferences and information on those planned for the future.

Delphi technique

Description

The Delphi technique is 'a method for structuring a group communication process so that the process is effective in allowing a group of individuals, as a whole, to deal with a complex problem' (Linstone and Turoff 1975:3). Furthermore, it is 'a method for the systematic solicitation and collation of judgments on a particular topic through a set of carefully designed sequential questionnaires interspersed with summarized information and feedback of opinions derived from earlier responses' (Delbecq et al. 1975:10). It is used most frequently to integrate the judgments of a group of experts. A key feature of this technique, however, is that the respondents do not meet and their responses may be anonymous. We still consider it to be a dialogue method, however, because 'conversation' between the parties occurs, even though it is not face-to-face.

Three separate groups of actors are generally involved:

1. Respondent group: those whose judgments are obtained through completing the process
2. Staff group: those who design the initial questions, summarise the responses and prepare the questions for subsequent phases
3. Decision-makers: those wishing to receive a product such as a consensus position from experts, or a recommendation (adapted from Delbecq et al. 1975).

Although some flexibility exists in implementation, the core method, as described by Delbecq et al. (1975:11), is as follows:

> First, the staff team in collaboration with decision makers develops an initial questionnaire and distributes it…to the respondent group. The respondents independently generate their ideas in answer to the first questionnaire and return it. The staff team then summarizes the responses to the first questionnaire and develops a feedback report along with the second set of questionnaires for the respondent group. Having received the feedback report, the respondents independently evaluate earlier responses. Respondents are asked to independently vote on priority ideas included in the second questionnaire and mail their responses back to the staff team. The staff team then develops a final summary and feedback report to the respondent group and decision makers.

Variations of this basic approach include:

- whether the respondent group is anonymous
- whether open-ended or structured questions are used to obtain information from the respondent group

- whether the responses are collected in written form or verbally—for example, over the phone
- how many iterations of questionnaires and feedback reports are used
- what decision rules are used to aggregate the judgments of the respondent group.

The number of participants can range from a few to many hundreds. The larger the number of iterations employed, the closer to consensus will be the result. Written questionnaires can be in pencil-and-paper form or distributed and returned using electronic communication tools including email and the internet. Computer-based systems, using highly structured questionnaires, can produce real-time findings.

Examples of its use in research integration

1. The environment: developing an environmental plan for a university

What was the context for the integration, what was the integration aiming to achieve and who was intended to benefit?

Senior administrators at Dalhousie University in Halifax, Nova Scotia, Canada, were aware of a significant gap between the university's environmental policies and their implementation. As a result, they resolved to develop an implementation plan that would be acceptable to all those who would be responsible for making it work.

Those responsible for developing the implementation plan used the Delphi technique

> to consult with key representatives of the university community in order to generate ideas about the most desirable and feasible ways in which to incorporate the new Environmental Policy into the activities and structure of the university…Modifying a Delphi study for policy research can be used to generate ideas and provide decision-makers with the strongest arguments for and against different resolutions to an issue. (Wright 2006:763)

Who did the integration, how was it undertaken and what was integrated?

A panel of 28 individuals was selected, with equal numbers drawn from the identified key stakeholders: 'students, staff, faculty, and administrators.' A core feature was that the Delphi study participants would be anonymous to one another, as the Delphi technique was implemented by email between the panellists and the project managers, rather than through face-to-face discussion. This was considered important as it gave equal weight to each panellist's judgments, avoiding problems that the power imbalances among the panellists (for example, between students and faculty) might otherwise create. No information was provided on what was integrated.

The Delphi questionnaires were distributed and responses received by email. Round one was the open-ended question: 'After reading the Environmental Policy, what recommendations do you have to incorporate it into the activities and structure of Dalhousie University?' A master list of 125 suggestions was developed from the responses. In round two, the participants were asked to review the master list from round one and rate each item for desirability and feasibility (separately) on a five-point Likert scale. The responses to the second round were analysed statistically for measures of central tendency and dispersion. The items were categorised as those that received consensus: a) for being desirable and feasible; b) for being desirable but not feasible; or c) rated as either not desirable or unsure. Each participant received a personalised questionnaire in round three, listing that person's ratings in round two, along with the group responses. They were asked to reconsider their ratings and make any changes. In round three, the majority of participants modified two to five of their round two ratings.

What was the outcome of the integration?

The results of the Delphi technique study were used by the university managers as a key input to developing the Environmental Policy Implementation Plan. The features that made the Dephi technique useful were identified as anonymity, encouraging exploratory thought and developing innovative ideas, achieving consensus, serving as an educative tool about environmental issues and being a tool for empowerment (Wright 2006).

2. Public health: estimating the incidence of *Salmonella* poisoning

What was the context for the integration?

Despite food poisoning through food-borne *Salmonella* infection being an important public health problem, in the United Kingdom in the mid 1990s, official statistics were not able to provide an accurate estimate of the incidence of infections. It was agreed that the official data significantly underestimated the true incidence, but experts' views differed about the level of underreporting.

What was the integration aiming to achieve and who was intended to benefit?

The Delphi technique was used by Henson (1997:197) to: a) 'reconcile differences in expert opinion and provide more reliable estimates of the incidence of food-borne *Salmonella*'; and b) to identify expert opinion about the effectiveness of the available measures for control of the infection. This dialogue method was chosen because it was, in the view of the person who implemented the study,

> a recognised technique for reconciling differences in group judgements where there is inherent uncertainty as to the actual state of the world.

In this case, the group consists of experts on food-borne *Salmonella* in the UK. The aim is to generate data which may overcome acknowledged problems with published statistics. (Henson 1997:196)

Who did the integration, how was it undertaken and what was integrated?

The Delphi study was initiated by conducting a workshop in which seven experts in food-borne *Salmonella* infection examined the issues to be covered in the survey. They did so using the nominal group technique, discussed below. The workshop identified the precise wording to be used in the Delphi study questions.

Some 62 experts (their areas of expertise were not specified) in food-borne *Salmonella* infection, identified by workshop participants, were then invited to be part of the Delphi study, and 42 of them agreed to do so. Five Delphi rounds were conducted during a seven-month period, with three exploring the experts' judgments of the incidence of infection and all of them exploring the effectiveness of control measures. This was done by means of questionnaires, but further details were not given.

The first question was: 'What would you estimate to be the total number of persons ill due to infection with non-typhoid *Salmonella* in the UK from all sources (food and non-food), over the course of one year?' The second asked what proportion of infections participants thought was food-borne and the third invited them to identify the proportion of cases by type of food. For each question, they were asked to advise how they produced their estimates and any difficulties they encountered in doing so. The results of the first and second rounds were fed back to participants, showing them the median, minimum and maximum responses from the whole panel and inviting them to revise their estimates of incidence.

In round one, participants were also asked to list the control strategies available for reducing the incidence of food-borne *Salmonella* infection. In round two, they were asked to refine the list and in round three the refined list was presented along with the question, 'Taking each control strategy in turn, consider how effective it would be at reducing the total incidence of food-borne non-*typhi Salmonella* in the UK?' This question was repeated in the fourth and fifth rounds, with the findings of the previous round fed back to participants.

What was the outcome of the integration?

An important outcome of the process was the narrowing of the range of estimates for the incidence of infection, as participants reflected on the median and range of responses to the incidence questions. Regarding the effectiveness of control measures, one approach (food irradiation) was identified by the panel as being particularly effective. Considerable disagreement remained, however, about which other measures were effective, even after three rounds considering this question. The author concludes that this is not really problematic as the Delphi study 'provides a good summary measure of expert opinion in an area which is characterised by great uncertainty' and 'the spread of responses provides a good indication of the range within which we can expect to find the actual state of the world' (Henson 1997:203).

3. Security: developing a new medical school curriculum addressing bio-terrorism

What was the context for the integration?

Since the 11 September 2001 attacks on the New York World Trade Centre and the Pentagon in the United States, responding to the medical sequelae of bio-terrorism and biological warfare incidents is no longer considered solely the province of emergency medicine specialists. Rather, it is seen as something that all healthcare providers need to be prepared to handle.

What was the integration aiming to achieve and who was intended to benefit?

Medical educators in the United States set out to develop new medical school curriculum guidelines relating to bio-terrorism so as to equip the next generation of medical graduates to be able to respond to this threat. They used an internet-based Delphi survey to identify the educational objectives to be covered by the curriculum guidelines (Coico et al. 2004).

The Delphi technique was chosen for this purpose because, in the views of those who wished to develop the new curriculum guidelines, it 'can provide a relatively rapid means of gaining a consensus on complex issues' (Coico et al. 2004:367).

What was being integrated?

This consensus came through the integration of the judgments of a group of experts in microbiology and immunology, who were engaged in medical education in US universities. Some 89 per cent of panellists had PhD degrees, 7 per cent were physicians and 77 per cent were involved in medical curriculum development. Two-thirds rated their expertise concerning bio-terrorism, biological warfare and bio-defence as 'strong' or 'moderate'.

Who did the integration and how was it undertaken?

A total of 237 people were invited, by email, to join the Delphi panel and 64 (25 per cent) participated in one or more rounds. The Delphi process comprised three internet-based rounds using 'a dynamic Web-based questionnaire' (Coico et al. 2004:367). The responses were captured from the web server onto spreadsheets. Before the first round, participants provided demographic information including self-assessment of their expertise in bio-terrorism.

Previous workshop discussions had produced a list of six content-related curriculum categories for bio-terrorism teaching and learning: general issues, bio-defence, public health, infection control, infectious diseases and 'weaponizable toxins' (Coico et al. 2004:368). These were put to the participants in the first Delphi round, and they were asked to add knowledge, skills and attitude objectives to the list of educational objectives. They were also asked for suggestions about any content areas that seemed irrelevant to the project. In round two, the responses to round one were fed back to participants and they were asked to assess, for three identified levels of medical training, the relative importance of each objective. The results of round two were fed back to the panel in round three in the form of percentage endorsed figures, and they were asked to identify their top five curriculum objectives in each category. They were also asked to rate the usefulness of nine different methods of teaching/learning and assessment of bio-terrorism and bio-defence topics.

The products of round two were also passed to an independent expert committee to obtain their views. This separate, independent committee had members who were experts from other professions and disciplines concerned with the issues being addressed by the panel. Its function was to receive the panel's findings and consider their implications.

What was the outcome of the integration?

Although the authors of the paper reporting on this project stated that they would have benefited from a higher participation rate, they felt that the Delphi technique 'provided an opportunity to explore bioterrorism-related curriculum issues in depth' (Coico et al. 2004:372). The outcome was the inclusion, in the US Medical Licensing Examination, of approximately one-third of the educational objectives identified through the Delphi study.

4. Technological innovation: developing professional association policies and practices for shifting from paper to electronic communications

What was the context for the integration?

The Institute of Electrical and Electronics Engineers (IEEE) describes itself as the world's leading professional association for the advancement of technology, and the largest, with more than 365 000 members in more than 150 nations (<http://www.ieee.org/web/aboutus/home/index.html>). In the late 1990s, it identified the need to establish policies and procedures governing its transition from hard copy to electronic communication and dissemination of information within the institute and beyond it. Indeed, in 1996, it adopted the slogan 'IEEE: networking the world'.

What was the integration aiming to achieve and who was intended to benefit?

The institute used the Delphi method to assess the benefits of and obstacles to its transition to electronic communications. It saw this method as a

> technique that is considered appropriate when the research purpose is to glean and synthesize expert opinion about complex issues and to identify recommendations for addressing them. The technique is frequently used in exploratory research and in efforts aimed at technological forecasting, including technological trajectories and the impacts of technological change...Use of the method in this research project has allowed the researchers an opportunity to pool a wide range of expert opinion in order to arrive at a series of focused predictions that may guide the IEEE's approach during this significant transition period. (Herkert and Nielsen 1998:80)

Who did the integration, how was it undertaken and what was integrated?

A pool of institute members—the exact number was not reported—was identified by the project managers and invited to participate in the study. They came from five areas: institute leadership and staff, institute technical activities representatives, institute regional activities representatives, customers and 'informed others'. Forty agreed to participate in the study and 30 provided demographic information and responded to round one. (It is not clear if the Delphi questionnaires were distributed and returned electronically or in 'pen-and-paper' format.) In round one, participants were asked to assess:

1. the potential contribution of electronic communication and information dissemination in fulfilling the institute's strategic planning goals and objectives that did not rely explicitly on the use of electronic media
2. the impact of electronic communication and information dissemination with respect to the five strategic planning goals and objectives that relied explicitly on the use of electronic media.

They were also invited to provide open-ended commentary on the benefits of and obstacles to the use of electronic media (Herkert and Nielsen 1998:82–4). A round one example question was:

> Products and Services Objective: Make all IEEE information products and databases of value to members available in electronic form as quickly as possible.
>
> > I agree; making products and databases available in electronic form as quickly as possible is a valuable objective.

I disagree; making products and databases available in electronic form as quickly as possible is not a valuable objective.

Discuss your answer in the space provided below.

In round two, the panellists were given a synthesis of the obstacles to the Institute's increasing reliance on electronic communication derived from round one and were asked to identify the 10 obstacles that each respondent considered most problematic. In round three, they were provided with a list of the 11 major obstacles identified in round two and asked what actions the Institute should take to benefit maximally from electronic communications while avoiding its potential pitfalls. Content analysis was conducted on the responses to round three to identify the path forward.

What was the outcome of the integration?

This application of the Delphi technique resulted in the IEEE identifying six key factors affecting the adoption and use of electronic media:

1. characteristics of the IEEE as technology initiator;
2. characteristics of the potential individual adopter;
3. characteristics of the potential organizational adopter;
4. characteristics of the technology;
5. outcomes, and
6. characteristics of the contextual environment. (Herkert and Nielsen 1998:95–6)

This finding, combined with a content analysis of the panellists' qualitative responses, enabled the investigators to develop a range of recommendations for consideration by the executives of the Institute to guide it in embracing electronic communication methods.

Commentary

The Delphi technique is generally implemented by means of pen-and-paper, email or web-based questionnaires, or by one-on-one interviewer–interviewee questionnaires. This means that it does not entail face-to-face dialogue. Instead, a 'conversation' occurs by means of responding to the questionnaires and sharing all the participants' responses, one with another. What are missing are the additional communication cues—verbal and non-verbal—that occur in face-to-face dialogue. Here, the focus is on the contents of the message, the real wording, rather than the other features that constitute human communication. Nonetheless, we classify it as a dialogue method in that the iterations in the process have features similar to two-way communication in the face-to-face situation.

All four case examples applied the Delphi technique reasonably closely to the 'ideal type'. The number of iterations used varied, with three of the examples utilising three rounds and another (the public health example) five rounds—an unusually large number. This also demonstrates how it can be implemented flexibly, depending on the topic, the participants, resource considerations, and so on.

Most commonly, the method is used with a group of peers: experts with relatively equal status, a more-or-less common knowledge base and a shared epistemology. This was the case with the public health and security examples, all the participants in which were experts. In contrast, the first example illustrated the participation of three stakeholder groups, none of which was particularly expert on the topic. The fourth example demonstrates the method's use among a fairly diverse range of participants.

As a method of research integration, it is especially useful for complex problems about which uncertainty exists and for which expert judgment is needed to deal with this uncertainty. The problems are typically multifaceted and demand insights derived from different types of knowledge, experience and information. This means that the problem being addressed needs to be tightly defined and the questionnaires must deal explicitly with boundary issues. It is a highly task-oriented process, seeking answers to a tightly defined problem.

As the examples illustrate, the method is highly adaptable in terms of its contents. We are aware of at least one example of the method being given a title that reflects the contents being judged—namely, the 'Ethical Delphi' (Millar et al. 2006, 2007). This is not so much a methodological variant as the application of the standard Delphi technique to a particular content area—in this case, concerns about ethics and values.

This method, unlike those discussed above, relies very much on the people (the research integrators) who manage the process to make the syntheses and judgments. They develop the questions, score the responses and identify the conclusions, their validity and reliability and their utility. This allows for strong focus on the task (in contrast to unfacilitated face-to-face group processes where the focus can be readily diverted).

Origins and genealogy

The Delphi technique had its origins in the early 1950s' Cold War between the United States and the Soviet Union, when the RAND Corporation was commissioned by the US Department of Defence 'to apply expert opinion to the selection, from the point of view of a Soviet strategic planner, of an optimal US industrial target system and to the estimation of the number of A-bombs required to reduce the munitions output by a prescribed amount' (Dalkey and Helmer

1963:458). Its title refers to the Delphic Oracle, reflecting the fact that it was originally used as a forecasting technique with respect to science and technology.

Since then, the Delphi technique has been used many thousands of times in diverse sectors addressing a huge array of questions. Although the originators did not situate it (in their 1963 paper) in any particular body of theory, subsequent scholars have attempted to do so. A wide range of traditions in Western philosophy has been invoked in this context, with one schema, presented by Mitroff and Turoff (1975), demonstrating that the method can be understood through the Lockean Inquiring System (the basis of much empirical science), the Leibnizian Inquiring System (the basis of much theoretical science), the Kantian Inquiring System (which combines both of these approaches) and the Singerian-Churchman Inquiring System.

Scholars have concluded that there is no single school of philosophy that best captures the theory underlying the Delphi technique (Mitroff and Turoff 1975).

Further reading on the Delphi technique

Adler, M. and Ziglio, E. (eds) 1996, *Gazing Into the Oracle: The Delphi method and its application to social policy and public health*, Jessica Kingsley Publishers, London.

Delbecq, A. L., Gustafson, D. H. and Van de Ven, A. H. 1975, *Group Techniques for Program Planning: A guide to nominal group and Delphi processes*, Management Application Series, Scott, Foresman, Glenview, Ill.

Linstone, H. A. and Turoff, M. (eds) 1975, *The Delphi Method: Techniques and applications*, Addison-Wesley Publishing Company Advanced Book Program, Reading, Mass.

Future search conference

Description

Future search conferences (and the related search conferences) have been conducted in various parts of the world since the late 1950s. They are large-group planning conferences, using face-to-face dialogue to develop plans, including the identification of action steps. They begin with a focus on visions and use these to guide the proposals for action.

The implementation methods vary. Some proponents of this technique argue for limiting the number of future search conference participants to about 60–80, all meeting in one room and in active dialogue, on the grounds that more than this number means that productive dialogue is not feasible. Others are comfortable with far more participants. In these instances, participants are broken up into smaller groups and meet in separate break-out rooms, coming together for plenary sessions at which experts provide inputs. At the plenary session reporters for the smaller groups also provide feedback to the other participants. The tasks of the conferences are expressed as 'The future of…'.

The Future Search Network has documented a number of conditions for the success of future search conferences. They include:

- getting the 'whole system' in the room; invite a significant cross-section of all parties with a stake in the outcomes of the conference
- exploring the 'whole elephant' before seeking to fix any part; get everyone talking about the same world; explore the global context before focusing on local issues
- emphasising common ground and future focuses, while treating problems and conflicts as information, not action items
- encouraging self-management and responsibility for action by participants before, during and after the conference (adapted from material at <http://www.futuresearch.net/>).

The Future Search Network provides an example of a typical three-day future search conference, with the following stages identified. The network recommends a three-day schedule, as the two intervening nights provide time for participants to process or reflect on the events of the day.

Day 1, afternoon. *Focus on the past*: people make timelines of key events in the world, their own lives and in the history of the future search topic. Small groups tell stories about each timeline and the implications of their stories for the work they have come to do.

Focus on present, external trends: the whole group makes a 'mind map' of trends affecting them now and identifies the trends that are most important for their topic.

Day 2, morning. *Focus on present, external trends (continued)*: stakeholder groups describe what they are doing now about key trends and what they want to do in the future.

Focus on the present: stakeholder groups report what they are proud of and sorry about in the way they are dealing with the future search topic.

Day 2, afternoon. *Ideal future scenarios:* small groups put themselves into the future and describe their visions—their preferred future—as if it has already been attained.

Identify common ground: small groups post themes they believe represent common ground for everyone.

Day 3, morning and early afternoon. *Confirm common ground:* the whole group meets to agree on common ground, integrating the diverse visions for the future elicited in the previous stages.

Action planning: volunteers sign up to implement action plans.

Future search conferences have been conducted in many different sectors, including commerce and industry, local communities, religious communities, schools and higher education, the environment, government, health care and human services. A recent example was the 'Bendigo+25 Future Search Conference', at which 'community members gathered to consider what kind of place Greater Bendigo would be in 2030. They worked together to develop a shared vision, values and key future directions' (City of Greater Bendigo n.d.).

Future search conferences are particularly useful in situations of rapid change (for example, changes in knowledge, society, technology, the environment) where various stakeholders can be expected to make different judgments about the implications of change for the future. Indeed, the first future search conference, held under the auspices of the Tavistock Institute (London) in 1959, provided an opportunity for two aircraft manufacturing companies that were merging to create Bristol-Siddley to integrate the different knowledge, technologies, practices and perceptions of the future held by the staff and managers of the two companies. The result was a new type of aircraft engine—one that is still in use today.

Because of its focus on the future, the future search conference has a strong emphasis on vision as part of the overall judgment. It helps clarify the visions of researchers and stakeholders and can draw them together into a shared vision for the future.

Example of its use in research integration

Public health: reducing the human and economic burden of Repetitive Strain Injury (RSI)

What was the context for the integration?

In the early 1990s, repetitive strain injury (RSI) was an important public health issue in many nations. It was characterised by an uncertain and ambiguous nature, the area was conflict ridden and many interconnected individual, organisational and societal phenomena were involved (Polanyi 2001).

What was the integration aiming to achieve and who was intended to benefit?

Polanyi's involvement in the field in Canada led him to conclude that

> the need [existed] for various stakeholders involved with RSI to meet in a nonadversarial setting in order to communicate effectively and safely with one another. There was a feeling that there exists unnecessary conflict and division between groups, which could be overcome by effective dialogue and increased collaboration. Future Search seemed to provide an inclusive forum that could stimulate dialogue among researchers, policy makers, practitioners, and injured workers to build the innovation and collaboration needed to better prevent and treat these injuries. (Polanyi 2001:468)

The specific aim of the conference, as articulated by the design team and consultants (see below), was '[t]o stimulate collaborative action to reduce the human and economic burden of upper limb musculoskeletal disorders'.

What was being integrated?

Stakeholder groups were identified (designers and engineers, employers, ergonomists, healthcare providers, health and safety professionals, government officials, injured workers and their advocates, labour representatives, researchers and the media, the provincial compensation board and private insurers). Lists of individuals to be invited were developed from within each stakeholder group with the aim of maximising the diversity of conference participants. It was the judgments of these diverse groups of participants that were being integrated.

Who did the integration and how was it undertaken?

A team comprising a range of stakeholders was established to design the future search conference and it was funded primarily by the (Ontario) Institute for Work and Health. It engaged two consultants experienced in implementing this intervention to have carriage of the project.

The conference had 56 participants and was held over three days in May 1998 following the standard, staged approach advocated by Weisbord and Janoff (2000):

- reviewing the past
- assessing the present
- developing future scenarios (shared visions)
- reality checking and action planning.

What was the outcome of the integration?

The future search conference was evaluated systematically (Polanyi 2001) using the grounded theory method of Corbin and Strauss (2008). (As is commonplace with the evaluation of the application of dialogue methods, the research integration processes and outcomes of the conference were not evaluated.) The evaluation revealed a high level of participant satisfaction with its process and outcomes. The conference succeeded well in building common ground between the participants, with agreement that

> (a) RSI is a 'real' concern for many and is having a serious impact both on people's lives and on economic costs, (b) RSI is complex and caused by several factors including individual behavior and conditions in the workplace, (c) all stakeholders need to work together to prevent and treat RSI through a systematic approach, and (d) further research is needed to identify and disseminate best prevention practices and treatments. (Polanyi 2001:473)

There was also agreement that action was needed on a number of fronts, including:

> (a) the identification and transfer of best practices in prevention, diagnosis, and treatment, and the establishment of the economic benefits of taking action to prevent RSI; (b) raising awareness through education and training based on what is known about the nature, scope, and effects of RSI; (c) promotion of a multistakeholder process through which all parties have input into solutions and accept shared responsibility for the problem, and (d) the provision of appropriate incentives for action, although this meant very different things for different people (Polanyi 2001, p. 473)

It was not possible, however, to reach common ground on all the issues. Four remained unresolved:

- which approach is needed: legislated standards or voluntary action?;
- do we know enough to act?,
- the relationships between productivity and worker health, and
- the right to pain-free work (Polanyi 2001).

The conference agreed on a range of actions to be taken and a follow-up conference was held to review the implementation and outcomes of the actions agreed on.

The establishment of common ground and a degree of shared visions for the future, and the preparation of actions plans that involved multi-stakeholder collaborations, were seen by many participants to be key achievements of the conference.

Commentary

This method contrasts with some of the others discussed in that the conferences are frequently very large, sometimes including hundreds of participants. This provides scope for great diversity, including significant opportunities for researchers' inputs to the process and shaping its outcomes. In the example provided—concerned with finding a shared approach to RSI in Canada—researchers were among the 56 conference participants providing an opportunity for research insights and products to be integrated with the inputs from other participants. This research-based knowledge came from a number of disciplines and was integrated with the knowledge, perspectives and visions of other stakeholders, including people adversely affected by RSI.

Unlike some other dialogue approaches, future search conferences focus explicitly on developing action plans for implementation after the conference concludes, action plans that reflect the shared understandings, visions and common ground established in the conference itself.

The method has been used in diverse settings. In terms of research integration specifically, one can envisage it being used in research institutions where disciplinary barriers need to be addressed to produce an integrated approach to a program of research, based on a set of agreed goals and action plans.

Origins and genealogy

Eric Trist and Fred Emery developed the search conference approach (the predecessor of the future search conference) at the Tavistock Institute in 1959. Others have subsequently modified the initial model, with Marvin Weisbord and Sandra Janoff particularly prominent in recent decades as developers and proponents of future search conferences. Bryson and Anderson (2000) and Oels (2002) explore the similarities and differences between search conferences and future search conferences. They include the selection of participants, with search conferences limiting participants to people with the capacity to implement action plans rather than the broad cross-section of stakeholders in future search conferences; grouping, with large groups dominating search conferences and a mix of small and large groups in future search conferences; and the methods of handling conflicts, with time spent discussing and clarifying differences in search conferences, whereas in future search conferences any disagreements are acknowledged but not discussed further.

Further reading on future search conference and search conference

Emery, M. and Purser, R. E. 1996, *The Search Conference: A powerful method for planning organizational change and community action*, Jossey-Bass Public Administration Series, Jossey-Bass, San Francisco, Calif.

Future Search Network 2003, *Future Search Network*, <http://www.futuresearch.net/>

Weisbord, M. R. (ed.) 1992, *Discovering Common Ground: How future search conferences bring people together to achieve breakthrough innovation, empowerment, shared vision, and collaborative action*, Berrett-Koehler, San Francisco.

Weisbord, M. R. and Janoff, S. 2000, *Future Search: An action guide to finding common ground in organizations and communities*, Second edition, Berrett-Koehler, San Francisco.

Most significant change technique

Description

The most significant change (MSC) technique is a relatively new dialogue method for monitoring and evaluating complex interventions. Its main focus is on program improvement. It contributes to organisational change and targeting of services/activities towards the attainment of valued outcomes. It is highly participatory and has at its core the generation, analysis and use of stories. The technique is also known as 'monitoring without indicators' and 'the story approach'.

Its main purpose is

> to facilitate program improvement by focusing the direction of work towards explicitly valued directions and away from less valued directions. MSC can also make a contribution to summative evaluation through both its process and its outputs. The technique involves a form of continuous values inquiry whereby designated groups of stakeholders search for significant program outcomes and then deliberate on the value of these outcomes in a systematic and transparent manner. (Dart and Davies 2003:137)

The most significant change technique involves 10 steps (six or seven in earlier descriptions):

1. Starting and raising interest
2. Defining the domains of change
3. Defining the reporting period
4. Collecting significant change stories
5. Selecting the most significant of the stories
6. Feeding back the results of the selection process
7. Verification of stories
8. Quantification
9. Secondary analysis and meta-monitoring
10. Revising the system (Davies and Dart 2005).

A small number (three to five) of loosely defined domains within which the stories are located are selected—for example, 'changes in the quality of life of the people affected by the program'. Stories are then generated by various stakeholders close to program implementation and knowledgeable about its outcomes within each domain using the question 'During the last month, in your opinion, what was the most significant change that took place in the program?'. In one application to rural extension in Australia (Dart and Davies 2003), this was expanded by asking 'What happened?', 'Why do you think this is a significant change?' and 'What difference has it made/will it make in the future?'.

The respondents—the producers of the stories—allocate their stories to a domain. It is useful to include as one of the domains 'lessons learned' as this tends to reduce the bias towards stories illustrating positive outcomes of the program. The stories are typically one to two pages in length.

Those managing the most significant change process and the program managers closest to the program implementation level (these may be the same people) then select the most significant of the stories and pass them up the organisational hierarchy. At each level, they are read, discussed and the most significant are selected. Feedback is given to the lower levels, particularly as to the reasons why individual stories have been accepted for passing up, or set aside. The process reduces a large number of stories considered important at the local level to a smaller set that are most important at a higher level within the organisation. The refined set is discussed by senior management or funding bodies and taken into account in subsequent strategy development.

Various techniques exist for verifying the stories and attaching quantitative indicators to them in situations where this is feasible and useful.

As a method for research integration, the most significant change technique can be applied to monitor and evaluate research integration in a wide variety of complex interventions, particularly when information on outcomes and the value base of the interventions are important. It has been pointed out that

> The types of programs that are not adequately catered for by orthodox approaches and [which] can gain considerable value from MSC include programs that are:
>
> - complex and produce diverse and emergent outcomes
> - large with numerous organisational layers
> - focused on social change
> - participatory in ethos
> - designed with repeated contact between field staff and participants
> - struggling with conventional monitoring systems
> - highly customised services to a small number of beneficiaries. (Davies and Dart 2005:12–13)

It is not as useful in situations in which the implementation processes and outcomes are straightforward and the causal paths connecting inputs and outcomes are clear. In these circumstances, more traditional quantitative indicators are often adequate.

One of the developers of the most significant change technique explains that 'MSC can be conceived as a form of dynamic values inquiry whereby designated groups of stakeholders continuously search for significant program outcomes and then deliberate on the value of these outcomes. This process contributes to both program improvement and judgment' (Dart and Davies 2003:140).

The most significant change technique is also useful at the following interfaces:

- research and policy—for example, assisting senior decision makers to understand their programs' outcomes and the values they reflect
- research and professional practice—for example, assisting professionals responsible for designing and implementing complex interventions to understand how they have been implemented and with what outcomes
- research and those affected by the research—for example, providing people's own descriptions and analyses of program implementation, outcomes and attribution of causality.

The most significant change technique is a useful dialogue technique when integration is desired across powerful and weak players—for example, high-level program managers or program funders, on the one hand, and field staff and people intended to be program beneficiaries, on the other.

Example of its use in research integration

Natural resource management: evaluating a multifaceted rural dairy extension project in Australia

What was the context for the integration, what was the integration aiming to achieve and who was intended to benefit?

Target 10 is an extension dairy project that operates in Victoria under the aegis of the Department of Natural Resources and Environment (<http://dairyextension.com.au/>). Target 10 has been operating since 1992 and has diverse stakeholders, including farmers, university researchers, government extension officers and industry groups. In the late 1990s, the key stakeholders used the most significant change technique as a component of the project's evaluation (Dart and Davies 2003).

Who did the integration and how was it undertaken?

The most significant change technique was applied in six steps:

1. pilot testing and familiarisation with the process
2. establishing the domains of change
3. establishing a reference group
4. establishing a method to collect and review the significant change stories
5. holding an annual round-table meeting for funders and others to review the stories
6. conducting a secondary analysis of all the stories generated by the project.

A simple form was developed to assist farmers and extension officers, in particular, to develop the stories. The Delphi technique was used with some 150 program stakeholders to develop the domains of change—namely, changes in on-farm practice, changes in profitability or productivity, changes in farmer decision-making skills and any other significant types of change. After the establishment of a reference group, the staff and others involved through program committees were asked to write stories on the forms provided for that purpose. Few were forthcoming so stories were also captured from verbal presentations at meetings and transcribed onto the forms.

Each regional committee reviewed the stories coming from the field and selected one from each of the four domains for consideration at the regular two-monthly or three-monthly state-level meetings. In addition, they documented why they had chosen those stories. The selection process involved the stories submitted being read out at regional meetings and voted on by the participants as to their usefulness. Since a great diversity of views occurred about which were most useful, the stories were discussed in detail with the aim of attaining group consensus, particularly about the value of the outcomes described.

There was a 12-month process of developing and discussing the stories: extension staff and other program people discussed with farmers their most significant change experiences and documented them, the regional committees discussed and selected the best stories and this process was repeated at the state level.

What was the outcome of the integration?

At the final stage, the funders and other key participants in the program met to share their reactions to the 24 stories that had filtered up to them from the regions. Those discussions revealed great diversity in key stakeholders' reactions to the stories: the process demonstrated that they did not have a shared vision as to what Target 10 was intended to achieve for the dairy farmers or the dairy industry more broadly. After discussion, they were able to agree on one story that portrayed the types of outcomes that they all wished to see from the program and were happy about funding. This story was about a dairy farmer who attended a series of Target 10 courses, implemented what had been learned and, as a result, markedly increased the efficiency of use of farm inputs and gained confidence to move to managing a larger dairy farm.

What was being integrated?

At one level, the experiences of a diverse range of dairy farmers were being integrated. This occurred through them, and the extension officers who worked closely with them, documenting their stories and sharing them with their peers and people higher in the Target 10 hierarchy. More significantly, however, was the integrative work that occurred through discussing and judging the stories in terms of what they revealed about the most valuable outcomes of the program. This was a process of integration that occurred step by step at different levels in the program, until the winnowing process enabled the top level of managers to integrate their judgments to identify a single narrative that best captured the program's intended outcomes.

This method integrated the experiences, and judgments of them, of a diverse range of dairy farmers. In this case, it was also instructive to examine how the method synthesised some of the key components of those judgments—namely, visions, values and interests.

The process revealed, at the regional and state levels, that the various stakeholders—including at the middle-management level and the top-level funding bodies—had widely diverging visions as to what Target 10 was established to achieve. They differed in the values they placed on the various program outcomes, and this had implications for the whole program's direction. Through intense discussion of the stories, and voting on them, their separate visions as to what the program should attain were revealed, compared and tested. The process, grounded in the real-life stories of what people in the program considered to be the most significant changes, assisted the key stakeholders to examine their individual visions and their perceptions of what outcomes were most valued to produce a common, integrated vision for the program as a whole. Agreeing on the one particular story that encapsulated their shared perceptions was an important step.

Interests were also integrated through this application of the most significant change technique. The key stakeholders—funders, senior officers of the agriculture department, scientists and other university-based people—brought different interests to Target 10. The dialogue process of examining the stories of change helped make these interests explicit: able to be examined, compared, contrasted and weighed. The result was the attainment of a degree of consensus on desired program outcomes—in other words, a movement towards accommodating differing interests.

Commentary

The core of this dialogue method is tapping the experiences and judgments of people involved at different levels of an intervention, with respect to its outcomes. The method seeks to identify the most significant changes that have occurred and to provide information on what are seen as the causal pathways that have produced the outcomes. In this manner, it integrates the evaluative judgments of various players, along with the underlying interests and values that feed into these judgments.

It applies democratic principles by giving voice to program participants at all levels, and providing feedback from the senior levels to lower levels about the program outcomes judged to be most worthy or desirable.

Our example of its use in the Target 10 dairy industry extension project illustrates researchers as one group of participants interacting through dialogue with others (for example, government funding body representatives, dairy farmers themselves, extension officers, and so on). It can operate, then, at a variety of interfaces between researchers and others, eliciting and judging the different perspectives of the various players.

While this is a strength as a component of a program evaluation, other evaluation techniques—not necessarily dialogic in nature—are also needed. These sometimes include quantitative performance indicators to supplement the most significant change technique's narrative approach, and the use of the most significant change stories (along with other information sources) to develop the program logic. On the other hand, a strength of the method is that it can provide usable performance information in the absence of quantitative performance measures. This is particularly useful in complex, rapidly changing situations where the implementation of the intervention gives rise to emergent properties.

The Target 10 example is another illustration of how dialogue methods can be combined. In this case, the organisers used the Delphi technique with some 150 program stakeholders to clarify the domains of the program of change that would be explored in the most significant change project.

Origins and genealogy

The most significant change technique was initially developed by Dr Rick Davies to contribute to the evaluation of a multifaceted social development project on Bangladesh (Davies 1996). It was developed further by Davies and Dr Jessica Dart (<www.clearhorizon.com.au/>). Dr Dart has applied the technique in a number of situations in Australia.

The method has its origins in evolutionary epistemology, 'a branch of epistemology that applies the concepts of biological evolution to the growth of human knowledge' (Wikipedia Contributors 2009). As Davies and Dart (2005:73) explain:

> [I]n cultural evolution, the meaning of a given event…may be interpreted in a variety of ways by people. Some of those interpretations may have a better fit with the world view of the people concerned, and thus become more prevalent than other views held in the past. Within this newly dominant view, further variations of interpretations may emerge, and so on.

> The MSC process…was an attempt to design a structured social process that embodied the three elements of the evolutionary algorithm: variation, selection and retention, reiterated through time.

Further reading on the most significant change technique

Dart, J. and Davies, R. 2003, 'A dialogical, story-based evaluation tool: the most significant change technique', *American Journal of Evaluation*, vol. 24, no. 2, pp. 137–55.

Davies, R. 1996, *An Evolutionary Approach to Facilitating Organisational Learning: An experiment by the Christian Commission for Development in Bangladesh*, Centre for Development Studies, Swansea, Wales.

Davies, R. and Dart, J. 2005, *The Most Significant Change (MSC) Technique: A guide to its use*, Rick Davies and Jess Dart, Trumpington, Cambridge, United Kingdom, and Hastings, Victoria, Australia.

Clear Horizon, <http://www.clearhorizon.com.au/site/index.htm>

Most significant changes, <http://groups.yahoo.com/group/MostSignificantChanges/> (includes a repository of files illustrating the application of the technique in 10 countries)

Nominal group technique

Description

The nominal group technique is used to assist participants in the process of pooling their knowledge and, particularly, their judgments to arrive at decisions that are acknowledged by participants as being a genuine product of the group dialogue process. Being highly structured, it facilitates participation by preventing the group from being dominated by particular individuals, as each contribution is of equal value.

The nominal group technique operates in four stages:

1. Generating ideas: each individual in the group silently generates ideas and writes them down
2. Recording ideas: group members engage in a round-robin feedback session to concisely record each idea
3. Discussing ideas: each recorded idea is then discussed to obtain clarification and evaluation
4. Voting on ideas: individuals vote privately on the ranking of the ideas, and the group decision is made based on these rankings (Dunham 1998).

The developers of the technique state that it is not designed for routine meetings or for negotiating or bargaining. Rather, its focus is '*judgemental* decision making' (Delbecq et al. 1975:5, emphasis in original):

> The central element of this situation is the lack of agreement or incomplete state of knowledge concerning either the nature of the problem or the components which must be included in a successful solution. As a result, heterogeneous group members must pool their judgments to invent or discover a satisfactory course of action. (Delbecq et al. 1975:5)

Its specific purposes have been described as follows:

*   To increase creativity and participation in group meetings involving problem-solving and/or fact-finding tasks
*   To develop or expand participants' perceptions of critical issues within defined problem areas
*   To identify priorities among selected issues within a problem area, considering the viewpoints of differently-oriented groups (Pfeiffer and Jones 1975).

The nominal group technique is taught and used widely in the context of group processes. As an integrative method, it is particularly useful for synthesising judgments where different types and extent of knowledge and/or a diversity of opinions exist on a problem or issue. Participants need to have a commitment to dialogue and a willingness to accept the outcomes of the group process, even if the outcomes do not match the position they initially brought to it.

This dialogue method can be applied in two phases in research integration aiming to find solutions to real-world problems: knowledge exploration ('a search for major conceptual frameworks and broad insights') and solution exploration ('the refinement of broad insights by specifying components which should be included in the solution program') (Delbecq et al. 1975:124–5). One implication of this staged approach is that each stage can call on different participant or resource person expertise, with skills in broad conceptualisation particularly useful in the early stage, and technical expertise, to identify solutions, in the later stage (Delbecq et al. 1975).

Examples of its use in research integration

1. The environment: assessing environmental studies and geography students' views about fieldwork

What was the integration aiming to achieve, who was intended to benefit and what was the context of the integration?

British academics used the nominal group technique to assess the perceptions of environmental studies and geography students about the fieldwork components of their courses, and to review the alternatives to fieldwork implemented when the 2001 foot and mouth disease epidemic disrupted access to agricultural areas in the United Kingdom (Cousin and Healey 2003; Fuller et al. 2003).

What was being integrated, who did the integration and how was it undertaken?

Thirty-three final-year students from five universities participated. All were enrolled in programs that had a fieldwork component that had been withdrawn and had previous experience with fieldwork as part of their university education. The five universities were selected to represent a number of types of environmental science and geography programs, including large and small university departments and old and new universities.

The five groups (one from each university) each had three to 10 participants who volunteered to be involved, having found out about the project via posters and email promotions on their campuses. The nominal group technique was applied systematically, as recommended by Delbecq et al. (1975), to enable all participants' voices to be heard (without domination by a small number of powerful individuals) and, concurrently, to attain group consensus. A single facilitator conducted all five groups to maintain consistency. The stimulus questions used were:

Q.1 In the light of any previous fieldwork experiences, how could fieldwork have made this unit: a) better, b) worse?

Q.2 What impact do you think the loss or withdrawal of fieldwork had on your experience of the unit and understanding of the subject?

The group responses were categorised under 12 types of educational objectives derived from the theoretical literature: experiential, interest, technical, analytical/research, specific subject knowledge, holistic/transferable, assessment/workload, financial/resource, environmental, time, teaching/module delivery and social/group dynamic.

What was the outcome of the integration?

A high level of consensus was observed across the five groups. The findings of this application of the nominal group technique were that the environmental studies and geography students found fieldwork an 'overwhelmingly positive experience', and the many reasons for this were made explicit. The negative aspects (for example, the time and expense involved) were also explicated. These findings enabled the educationalists involved to draw conclusions and recommendations about the fieldwork components of their courses. In their papers, they have thoroughly documented how they implemented the nominal group technique and have discussed its strengths and weaknesses.

2. Public health: developing criteria to assess the appropriateness of innovative services in community pharmacy

What was the context for the integration?

Although community pharmacies are accepted as an important component of the mix of healthcare services, much variation exists in the amount, nature and quality of advice provided to customers by pharmacy sales staff about non-prescription medicines and the treatment of minor ailments.

What was the integration aiming to achieve, who was intended to benefit and what was being integrated?

The nominal group technique was used in the United Kingdom to develop criteria to assess the appropriateness (or otherwise) of pharmacy counter staff providing advice to their customers (Bissell et al. 2000). In this case, the technique was used to make explicit the knowledge of a group of experts and to synthesise their judgments on the criteria for assessing the appropriateness of advice provided by pharmacy staff to the public. The new understandings derived from the process could then be used for integrating the expert knowledge of people with experience in the pharmacy setting with academics' skills in developing assessment criteria and the workforce educational interventions flowing from their availability.

Who did the integration and how was it undertaken?

The first stage of the study entailed capturing information on advice giving by counter staff in 10 community pharmacies by means of audio recording, supplemented by direct observation. Stage two entailed using the nominal group technique to elicit expert opinion on the criteria to be used to judge the appropriateness of the advice given, owing to the absence of previous research and theory in the area. Participants were selected using a nomination approach: the researchers contacted people who they believed could identify who was most expert in the community pharmacy field. Ten people were nominated and invited to participate, and eight of these accepted the invitation. The stimulus question was: 'How would you assess whether a consultation between pharmacy staff and a consumer was (in)appropriate?'

The group identified 73 individual items considered important in answering the stimulus question and, through discussion, these were condensed into just 10 core criteria:

- the overall layout of the pharmacy
- the overall organisation of the pharmacy
- general communication skills
- what information is gathered by pharmacy staff
- how information is gathered by the pharmacy staff
- issues to be considered by pharmacy staff before giving advice
- rational content of advice given by pharmacy staff
- how the advice is given
- rational product choice made by pharmacy staff
- referral (Bissell et al. 2000).

The group determined that prioritising these was neither feasible nor necessary, and that all should be weighted equally.

What was the outcome of the integration?

Subsequently, the criteria developed through the nominal group technique were subjected to statistical analysis of their validity and reliability. The authors state:

> The developed criteria will allow us to identify dimensions of both appropriate and inappropriate advice provided in community pharmacies and provide the basis for education and training initiatives identified as a result of the research. In addition, we suggest that this research is highly relevant to informing the content, structure and operationalisation of protocols and/or guidelines associated with the management of minor ailments and the sale of medicines through community pharmacies. (Bissell et al. 2000:359)

The authors stated that the core advantage of the nominal group technique for their purposes was that it removed the potential for bias derived from professional hierarchies that might occur in less-structured group interactions. They also pointed to some of the technique's limitations, especially (in this case) the potential bias derived from the methods used to select the participants and the small number of participants.

Commentary

These examples show how the nominal group technique combines some of the advantages of the Delphi technique and of face-to-face group interaction. Like the Delphi technique, it is structured in such a way as to give each participant an equal say, hence avoiding the power differentials that often impact on face-to-face group decision-making processes. In addition, operating face-to-face means that participants have opportunities, through verbal and non-verbal communication channels, to better understand the judgments expressed by other participants.

The examples also illustrate how this method is particularly useful in situations where decisions need to be made, but information is missing, uncertainty exists and judgments are required. The pooling of knowledge and ideas and the sharing of judgments produce an integrated product. These examples are limited, however, in that they do not illustrate clearly integration between disciplines and stakeholders as their focuses are specific, tightly defined participants. In the first example, these were stakeholders (students potentially affected by the decisions that would be made based on the outcomes of the nominal groups) and in the second a group of experts judging competing criteria of the appropriateness of a service to the community.

Like the Delphi technique, the method tends to be restricted in the range of stakeholders involved owing to the need for them to have a degree of common knowledge and background. On the other hand, the Delphi technique and the nominal group technique are frequently used when other stakeholders—for example, decision makers—want answers to specific questions. In these cases, the interface between the (expert) participants and the users of the technique's findings are clear, as are the modes of product utilisation. The method requires a fair degree of common epistemology among the participants, along with willingness to listen, openness to new ideas and a commitment to compromise and find consensus.

The success of the nominal group technique is dependent, in part, on having a skilled facilitator to assist participants to discuss the ideas generated and the explanations thereof. This is a research integration role, one that can be played either by a trained researcher or a professional facilitator well briefed on the issue being explored.

It works well when a range of disciplinary insights is to be integrated, as (unlike in the case of the Delphi technique) opportunities are provided for participants to explain, discuss and justify their ideas, revealing their sources in particular disciplinary perspectives.

Origins and genealogy

Andre L. Delbecq (from the University of Wisconsin, Madison) and Andrew H. Van de Ven (from Kent State University) developed the nominal group technique in 1968. They advise that '[i]t was derived from social-psychological studies of decision conferences, management-science studies of aggregating group judgments, and social-work studies of problems surrounding citizen participation in program planning' (Delbecq et al. 1975:7–8).

Further reading on the nominal group technique

Delbecq, A. L., Gustafson, D. H. and Van de Ven, A. H. 1975, *Group Techniques for Program Planning: A guide to nominal group and Delphi processes*, Management Application Series, Scott, Foresman, Glenview, Ill.

Dunham, R. B. 1998, *Nominal Group Technique: A users' guide*, <http://courses.bus.wisc.edu/rdunham/EMBA/Fall_2006_Readings/TEAMS/dunham_ngt.pdf/>

Open space technology

Description

Open space technology (also known as open space dialogue) has been used for two decades as a method for facilitating dialogue among people keen to focus on an issue that is important for them, but for which the way forward is unclear. Its practitioners have demonstrated that, given commitment to the issues and acceptance of the process, open space technology can assist groups of people to identify and explore issues, identify opportunities for change and identify and set priorities among action steps to achieve desired goals. The approach is based on the theories of complexity, self-organisation and open systems (Heft n.d.).

Open space technology sessions can be less than one day long or continue for up to five days. The number of participants can be small (as few as five) or up to 2000 (Heft n.d.). They typically conclude with a written report, often produced before the final session, and frequently have follow-up communication activities between participants, such as blogs, email lists, and so on. One-day sessions, however, generally do not produce a report. Two-day sessions usually use the first day for intense discussions and the second day for report preparation. Three-day sessions usually have intense discussion on the first day, report preparation on the second and close attention to priorities and action plans on the third (Owen 1997a).

The facilitator of an open space technology session invites people to participate—people who are thought to be passionate about the topic and willing to work collaboratively with others on it. No agenda is prepared; instead, just a notice as to the topic or issue to be worked on.

The key elements of the setting are one or more circles of chairs, with circles being seen as the 'fundamental geometry of human communication' (Owen 1997b:5). The room has to be large enough to have a number of small groups sit in circles, and to change the configuration of circles as the process unfolds. Larger sessions will have break-out rooms. There also has to be ample blank wall space. Ideally, there is little else to clutter the space.

The facilitator provides a welcome and outlines the purpose of the activity: the themes or issues to be addressed. The four principles of the process are explained:

- whoever comes is the right people
- whatever happens is the only thing that could have
- whenever it starts is the right time
- when it's over, it's over (Owen 1997b:95).

The single law, the 'Law of Two Feet' (or the Law of Mobility) is explained: 'If, during the course of the gathering, any person finds him or herself in a situation where they are neither learning nor contributing, they must use their two feet and go to some more productive place'. The 'final admonition' presented is 'be prepared to be surprised' (Owen 1997b:98, 101).

The facilitator then asks people to come to the centre of the circle, introduce themselves and tell everyone else of one aspect of the session's theme about which they are passionate. The person records this aspect on a large sheet of paper along with their name and a suggested time and place for discussing the issue, before attaching the sheet to the wall. In this way, a first-draft agenda evolves. Owen (1997a) advises that the number of issues raised is usually about 30 for a group of 25–50 people and about 75 for a group of 100–200 people. Larger groups tend not to generate many more issues.

When all the issues for discussion have been posted, participants sign up for the groups with which they wish to be involved. At this stage, topics can be combined. Once this step is completed, the facilitator announces that she or he is departing, inviting the groups to get to work. The participants then gather in circles around designated topics. As each 75-minute discussion ends, people move to the next group of interest, and this process is repeated throughout the assigned period. The groups do not have facilitators; they run themselves. A group participant takes notes on each session and enters them into a computer at the end of the session. The reports from each session are progressively posted on the walls throughout the day and collated to create the open space report.

The whole group reassembles twice a day: in the morning for announcements and in the evening for 'news', as Owen (1997a) puts it. The afternoon session includes reflection on the day's activities. Before the final session, individuals usually take responsibility for follow-up activity, be it communication or action oriented.

Whether the judgments integrated by this method are expert or lay or a combination depends on the participants. Harrison Owen, the method's originator, states that 'Open Space Technology is effective in situations where a diverse group of people must deal with complex and potentially conflicting material in innovative and productive ways. It is particularly powerful when nobody knows the answer and the ongoing participation of a number of people is required to deal with the questions' (Owen 1997b:15). The process requires participants to shed their power roles in organisational hierarchies and interact as equals.

Example of its use in research integration

Public health: generating ideas and plans for the development of the United Kingdom's public health workforce

What was the integration aiming to achieve and who was intended to benefit?

Open space technology was used to integrate the judgments of a diverse group of public health practitioners and educators concerned about a large, complex domain of professional activity: public health workforce development in the United Kingdom (Brocklehurst et al. 2005:996). A national event was conducted at the University of West England to explore the ambitious, two-part question: 'In developing the public health practitioner workforce in England, what is needed, and how do we do it?'

Who did the integration and how was it undertaken?

In all, 34 people from 25 public health and education organisations participated. They were selected by the organisers from their own professional networks, using 'a mix of purposive and convenience sampling'. After the opening session, at which the two linked questions were posed, the participants identified and conducted 16 different discussion sessions. A written report came from each. Some of the sessions covered quite specific topics while others were more expansive. The themes included:

- Identifying skills practitioners need to help achieve national obesity targets
- Evolving community pharmacists into medicines managers and public health practitioners
- Assessing the impact of new national NHS pay and conditions strategy on emerging public health roles in primary care organizations
- Managing the tensions between increasing public choice and implementing potentially restrictive public health policies (such as smoking bans in public places)
- Developing public health practice beyond the National Health Service
- Liberating the minds of those who are 'supposed' to be public health practitioners but who appear resistant to broadening their role (Brocklehurst et al. 2005:997).

These are all topics where facts alone are not sufficient, but where judgment is needed to make progress. The self-selected discussion sessions linked like-minded people to develop a shared set of judgments about the current workforce situation, the options for progress and the most appropriate ways to move forward. The participants worked at their own paces to understand each other's positions and judgments, to explore them and to find common ground.

What was being integrated?

As noted above, the participants were selected through purposive and convenience sampling, drawing on the professional networks of the organisers of the event. The participants came from public health and education organisations, and from the public and private sectors. Further information was not provided.

What was the context for the integration?

In the opinion of some, government policy in the United Kingdom 'has stimulated something of a renaissance in public health' (Brocklehurst et al. 2005:996). While workforce development has been one part of this 'renaissance', relatively little has been done to increase the capacity and capability of frontline public health practitioners such as public health nurses and environmental health officers. A public health workforce development agenda would be one response to these needs, and the open space technology event was conducted as a contribution to developing it.

What was the outcome of the integration?

Two prominent outcomes of the open space event were identified:

1. participants subsequently convened a number of local and national workshops, involving open space event participants and others, focusing on the development of public health practice
2. the event organisers developed a simple conceptual framework for public health workforce development, based on the summary of proceedings from the open space event, and used this in decision making on follow-up workforce development strategy activities.

The integration of the judgments of experts that occurred through this process provided the basis for subsequent action to improve public health practice.

Commentary

Open space technology contrasts with the dialogue methods discussed above in being far less structured. The overarching topic is set in advance, the subtopics are generated by the whole group and free-flowing group processes operate from that point on. This means that all the participants have the opportunity—indeed, are encouraged—to set the agenda by defining the issues to be worked on. The underlying philosophy is that synthesis will occur through small-group discussions among people self-selected to address a topic about which they are passionate. Integration is taken to be an emergent property of the group process.

The method is used when the nature of a problem is reasonably clear, but uncertainty exists about the directions in which to travel to address the problem, and the specific action steps to be taken. Diversity among the open space participants is a feature as it enhances the likelihood that viable, acceptable options will be identified and, through discussion leading to integration of judgments, agreed on. This means that participants can be experts, lay people or a combination of both. The opportunity exists for a range of stakeholders, including researchers, decision makers and members of affected communities, to be involved.

Our example from public health illustrates another of the system's features—namely, the implementation, after the open space event, of the action steps agreed on there, including active communication of its findings to various stakeholders and active, goal-oriented engagement with them.

Although the example provided illustrates well the standard application of the method, it does not go far in illustrating its utility for the specific purpose of research integration. One could readily envisage the topic being addressed by means of the open space technique, focusing on research integration, if the participants were public health workforce researchers and workforce managers who needed to make decisions about the future of the workforce.

Origins and genealogy

Harrison Owen organised an international conference of 250 participants in 1983 and was most frustrated by the experience, feeling that the benefits did not justify the effort involved in organising and running the conference. He was struck by the fact that everyone enjoyed one aspect of the conference: the coffee breaks. He determined to develop a dialogue tool that combined 'the level of synergy and excitement present in a good coffee break with the substantive activity and results characteristic of a good meeting' (Owen 1997b:3).

Owen ran his first open space conference in 1985 and, since then, especially with the publication of the first edition of his users' guide in 1992, the technology has been taken up and applied extensively. Open space technology has been used in commerce, government and community settings across the world, including a World Bank-sponsored Youth Open Space Dialogue on the topic 'How Do We Create a Better Future—The issues and the opportunities', held in Singapore in September 2006 over two days, with some 230 participants aged sixteen to twenty-four years (<http://www.worldbank.org/>).

Further reading on open space technology

Heft, L. n.d., *A Description of Open Space Technology*,
  <http://www.openingspace.net/
  openSpaceTechnology_method_DescriptionOpenSpaceTechnology.shtml>

Open Space World, <http://www.openspaceworld.org/>

Owen, H. n.d., *A Brief Users' Guide to Open Space Technology*,
<http://www.openspaceworld.com/users_guide.htm>

Owen, H. 1997b, *Open Space Technology: A user's guide*, Second edition,
Berrett-Koehler Publishers, San Francisco.

Owen, H. (ed.) 1995, *Tales From Open Space*, Abbott Pub., Potomac, Md (case
studies of the application of open space technology; full text online at
<http://www.openspaceworld.com/Tales.pdf>).

Scenario planning

Description

Scenario planning—also known as scenario thinking or scenario testing—is a dialogue method for the integration of judgments that is widely used in business, military and government settings. It has a particular emphasis on dealing with uncertainties, specifically responding to the need of many organisations to plan for uncertain futures. Typically, the judgments of experts are integrated through this process.

Although there are several different 'schools' of approaches to scenario planning (Bradfield et al. 2005), we provide here a generic description based on those that emphasise dialogue. (Other approaches use computer modelling and/or scenario development by outside experts, rather than group dialogue processes.)

The method has been described in the following terms:

> Scenarios are a way of developing alternative futures based on different combinations of assumptions, facts and trends...They are called 'scenarios' because they are like 'scenes' in the theater—a series of differing views or presentations of the same general topic. Once you see several scenarios at the same time, you better understand your options or possibilities. (Caldwell n.d.)

The goal of scenario planning is not to predict, but to gain foresight. It responds to the wise statement attributed to the French diplomat Talleyrand, 'When it is urgent, it is already too late' (quoted in De Jouvenel 2000:39), and works at the boundaries of knowledge. Scenario planning provides information to assist planners as they contemplate the forces that will shape their organisations and their performance in the future and how to be in a position to deal with, or benefit from, those forces in a proactive, rather than reactive, manner.

One authority (Caldwell n.d., drawing on Schwartz 1991) has identified eight steps in scenario planning:

1. identify the focal issue or decision
2. identify key forces in the local environment
3. identify driving forces
4. rank by importance and uncertainty
5. select scenario logic
6. flesh out the scenarios
7. identify implications
8. select the leading indicators and signposts.

Using scenarios involves a shift from defining managerial competence as knowing where we are now, where we will be in the future and having a clear path towards goal attainment, to accepting that we cannot know the future but, nonetheless, need to address uncertainties about the future as part of managerial processes (Wilson 2000).

An important issue in scenario planning is translating the products of the exercise into strategic decision making as 'experience shows that actually using scenarios for this purpose turns out to be a more perplexing problem than the scenario development process itself. As in the larger domain of strategy, implementation…turns out to be the crucial issue' (Wilson 2000:24).

Thus, scenario planning has two aspects: the development of the scenario and the application of the scenario. As a dialogue method for integrating judgments, scenario planning is particularly useful in highlighting uncertainty. By combining known information about the present and, where relevant, the past with understandings and assumptions about future change, uncertainty can be addressed and harnessed for planning purposes. With its focus on organisational learning (rather than individual learning), it has a useful role where research and organisations interface, including the research–business and research–policy interfaces. Importantly, the locus of integration is beyond the research sector, with research products being among the inputs to the scenario planning process.

This method is particularly useful for integrating the judgments of several people to improve understanding, through the development of the scenarios, as well as assisting decision makers with those integrated judgments, through the application of the scenarios. The method enables decision makers, be they in policy, business or other areas, to test likely consequences of alternative actions and, in some circumstances, it can enable costly real-life failures to be avoided.

Examples of its use in research integration

1. Natural resource management: identifying possible futures for the Austrian food supply chain

What was the integration aiming to achieve and who was intended to benefit?

Scenario planning methods were used to identify possible futures for the Austrian food supply chain and the related driving forces for landscape change out to the year 2020 (Penker and Wytrzens 2005). The Austrian Federal Ministry for Education, Science and Culture chose the method because ministry planners were aware of the high levels of uncertainty and the lack of quantitative data available in these areas. They acknowledged that '[t]he main information available to deal with this uncertainty is the personal judgement of practitioners and experts within the food chain itself' (Penker and Wytrzens 2005:176). Accordingly, a scenario planning exercise was conducted to integrate the judgments of the various actors involved in food systems and landscape planning to produce well-grounded future scenarios.

Who did the integration, how was it undertaken and what was integrated?

Those responsible for the exercise first developed a conceptual model describing the interactions between Austrian society, the food supply chain and the landscape. The key agents in the model were agricultural producers, food processors, food wholesalers and retailers, and consumers. Two scenario planning workshops were conducted: 1) a one-day workshop involving 25 practitioners from these sectors; and 2) another involving nine scientists from various disciplines who met five times and engaged in a subsequent email conversation.

The practitioners' workshop was conducted in May 2002 and had the title 'The Austrian Food Chain in 2020 and its Landscape Impacts'. The dialogue techniques applied by the workshop facilitators 'endeavoured to structure and organise individual ideas, to contrast and discuss divergent statements, to stimulate the imagination of those involved, and to improve logical consistency' (Penker and Wytrzens 2005:179). The plenary and small-group discussions followed a number of steps:

- step one: identification of the relevant driving forces
- step two: weighting of the driving forces regarding their importance
- step three: formulation of long-term development options for each driving force
- step four: generation of two coherent scenarios
- step five: naming of the two scenarios
- step six: analysis of consequences and strategies
- final debate
- feedback and end.

The scientists' scenario planning exercise occurred subsequently. It was undertaken because limited attention had been paid to the scientific evidence in the practitioners' workshop.

Each group developed two intentionally divergent scenarios, recognising that the likely future would be somewhere between the two. The practitioners' first scenario was the 'liberal market scenario', in which international trade agreements were liberalised, European Union agricultural subsidies reduced, farmers were seen as destroyers of the environment and consumers' food purchasing behaviour was driven largely by marketing and price. Their second scenario was labelled the 'protective policy scenario', in which protectionism became more prominent in agricultural policies, animal welfare and environmental protection became more significant and small business was important in food marketing. The scientists' scenarios were a 'fast world scenario' and a 'slow world scenario', which emphasised the speed of change in agriculture and food policies and practices, along with different degrees of state control.

Integration occurred through the facilitated workshop discussions in which participants discussed their individual weightings of the forces for change and reached group agreement on them, enabling the participants to move to the next steps in the dialogue, covering options for the future, refining these to two contrasting scenarios and teasing out the implications of the scenarios.

What was the outcome of the integration?

The facilitators of the process concluded:

> The scenario technique used [was] found to be a useful means of gathering and structuring disperse [*sic*] expert knowledge…scenarios can deal with uncertainty concerning the socio-economic driving forces of landscape change and therefore can be used as a preliminary step in formulating robust strategies for landscape management. (Penker and Wytrzens 2005:175)

It integrated the judgments of the practitioners to produce future scenarios but, importantly, needed a separate scenario planning exercise to make judgments about the applicability of the scientific data that the practitioners failed to address.

What was the context for the integration?

The scenario planning activity described here was a discrete module within a larger, interdisciplinary project called 'Fast Food–Slow Food: Food chain management and cultural landscapes'. The overall project was concerned with the impacts on the landscape of various patterns of food production and consumption, and aimed to identify sustainable ways of managing food chains.

2. Business: understanding the future of the international airline industry

What was the integration aiming to achieve and who was intended to benefit?

In the mid 1990s, British Airways realised that it needed to be better prepared to identify and interpret major changes in the environment in which it operated. It therefore initiated its first scenario planning activity in 1994 (Moyer 1996). The purpose was to ascertain and integrate the judgments of a large number of disparate experts within and outside the company about possible futures for the international airline industry and how British Airways should be positioned to benefit from the changes.

What was being integrated?

The process was initiated by the company's chief economist and implemented by a development team of eight staff from its Corporate Strategy, Government Affairs and Marketing Departments, along with an external consultant. A series of workshops to discuss the scenarios and their implications involved the directors and senior managers responsible for implementing the company's business plan, and key customers. Further details were not specified.

Who did the integration and how was it undertaken?

The senior management of the airline supported the initiative as an experiment: to see if it made a useful contribution to the company's strategic planning. It was implemented in two phases: first, scenario development, and then scenario workshops. Scenario development was conducted through interviews and the analysis of interview data. A team of eight British Airways staff conducted individual interviews with more than 40 senior company personnel, five group interviews with staff specialists, as well as a small number of interviews with people outside the company.

Two scenarios (known as 'stories') were written by the development team members and documented in booklets and presentations. They were the 'wild gardens' scenario, in which 'global integration goes so far that it is impossible to build lasting new structures of governance to replace the old, crumbling structures', and the 'new structures' scenario, in which 'shared values and new ways of organizing are found which enable growth to continue in a manageable, rather than socially disruptive, way' (Moyer 1996:174). In this case, the integrative activity to develop the scenarios was undertaken by the eight-person scenario development team.

Phase two of the project involved the development team distributing the stories in written form and conducting workshops with British Airways staff throughout the company. In all, 28 workshops were undertaken, targeting senior managers responsible for major components of the company's business plan. The purpose was to provide opportunities for managers to learn about and discuss the scenarios and then to generate new ideas for strategic planning. Brainstorming the implications of the scenarios and other creative techniques (not detailed in the case report) were used to integrate the insights from the scenarios developed with the existing knowledge and judgments of the managers.

What was the context for the integration?

The international airline industry was severely disrupted by the recession after the first Gulf War, with many major airlines experiencing financial crises. This highlighted to senior management the vulnerability of British Airways to global economic cycles and the need to be prepared for future changes in the business environment in which it operated.

What was the outcome of the integration?

Individual business groups within the company used the processes and products of the scenario planning workshops to modify their business plans. The processes stimulated dialogue between various parts of the company, management and the trade unions, increasing participants' understanding of the external influences on the company.

The senior managers of British Airways were pleased with the exercise, agreed to incorporate the scenarios generated into the company's strategic planning processes and to review, some time in the future, the value of developing new scenarios.

Commentary

Scenario planning contrasts with most of the tools discussed above in that it is applied most commonly in the business and policy sectors, rather than being used to address broad issues of concern to ordinary citizens. Typically, experts working inside these sectors use the method. They do not necessarily engage with researchers, and certainly do not normally engage with lay members of the community. Research products (data, information, knowledge, and so on) are, however, key inputs to scenario planning.

This is another approach designed explicitly to deal with uncertainty. While its focus is on organisational learning, interesting issues exist with respect to organisations' *use* of its products. Its products are scenarios that reflect the judgment of participants in the scenario planning exercises, but how these are used in decision making, or whether they are used at all, is an issue. The British Airways example is one in which the scenarios were carefully developed using research evidence and dialogue processes with stakeholders, and the products were passed on to senior management, but no evidence exists that they were used instrumentally in decision making.

The Austrian food-sector example is also informative in the context of research integration: the organisers of the exercise felt that the scenarios developed by the expert industry representatives inadequately incorporated scientific knowledge—research products—meaning that a second round of scenario planning was undertaken among scientists alone. This draws attention to the importance of designing scenario planning exercises, and facilitating the dialogue processes, in such a manner as to give due weight to the body of research evidence and to integrate it into the scenarios along with the judgments of participants.

Origins and genealogy

This dialogue method was first documented by nineteenth-century Prussian military planners with the approach being formalised by analysts at the RAND Corporation, the Shell Group and the French company SEMA in the 1960s and 1970s. Bradfield et al. (2005) provide details of the origins of the three main categories of techniques developed by these groups and their current manifestations.

Further reading on scenario planning

Bradfield, R., Wright, G., Burt, G., Cairns, G. and Van Der Heijden, K. 2005, 'The origins and evolution of scenario techniques in long range business planning', *Futures*, vol. 37, no. 8, pp. 795–812.

Caldwell, R. L. n.d., *Scenarios, Foresight and Change. Tutorial 2: Building scenarios*, University of Arizona, <http://ag.arizona.edu/futures/tou/tut2-buildscenarios.html>

Fahey, L. and Randall, R. M. (eds) 1998, *Learning From the Future: Competitive foresight scenarios*, Wiley, New York.

ScenarioThinking.org, <http://www.scenariothinking.org/>

Soft systems methodology

Description

Peter Checkland, the primary exponent of soft systems methodology, recently described it as

> an organized, flexible process for dealing with situations which someone sees as problematical, situations which call for action to be taken to improve them, to make them more acceptable, less full of tensions and unanswered questions. The 'process' referred to is an organized process of thinking your way to taking sensible 'action to improve' the situation; and, finally, it is a process based on a particular body of ideas, namely *system* ideas. (Checkland and Poulter 2006:4)

It is usually implemented in groups. The contributions of expert facilitators can be beneficial, but are not essential once participants understand the techniques. It is also possible that 'a researcher can be used as an intermediary, interviewing people and ensuring that each stakeholder is exposed to other perspectives' (McDonald et al. 2005:39–40).

The core of the method is to integrate judgments by treating purposeful action as a system, an adaptive whole. Changing one part of the system (initiating one course of action) will create changes elsewhere in the system. What particularly distinguishes the approach is that it reveals and deals explicitly with the potentially differing world views of the participants, examining how these world views (*Weltanschauung*) underlie their judgments. Soft systems methodology seeks *accommodation* among different, sometimes conflicting, world views.

The key characteristics and seven implementation stages of soft systems methodology are listed below. Checkland and other practitioners emphasise, however, that it is not a mechanical, linear process. Rather, it is inherently iterative. It moves from finding out about a problematic situation to taking action in the situation, and does so by carrying out some organised, explicit systems thinking about the real world:

1. Workshop participants express their perceptions of the problematical situation in an unstructured form.
2. They then develop a 'rich picture', a visual representation of the situation in which people find themselves. This generally takes the form of drawings and connecting lines on sheets of paper, providing a kind of a map of the real world and its challenges.
3. Some human activity systems relevant to the situation are carefully named in 'root definitions'. The aim is to produce common understanding and agreement among participants with respect to each system. These explicitly name a number of features of the relevant systems, and test them, using

the acronym CATWOE, standing for customers, actors, transformation process, world view, owners and environmental constraints, as follows:

- customers: those who might be helped or harmed by action
- actors: those who could be involved in making the system work
- transformation process: identifying the 'raw material' that the system will transform into 'end products'
- world view: the world view underlying people's desire to create the transformation
- owners: those with the power to stop the system from working
- environmental constraints: the things that have to be taken as given.

Each root definition makes plain its world view—that is, the point of view from which the (human activity) system is described, since one person's 'terrorist' is another's 'freedom fighter' (the example Checkland uses most frequently).

4. Conceptual models of the systems named in the root definitions are built. They are models of purposeful activity considered relevant to debate and argument about the problematical situation. They are not at this stage thought of as practical designs. They usually take the form of a map of the activities needed to make the system operational. Activities are listed and their relationships made explicit.

5. The debate about the situation is structured by comparing models with perceptions of the real world: the initial rich pictures. The aim of the debate is to find some possible changes that meet two criteria: systemically desirable and culturally feasible in the particular situation in question.

6. An action plan is developed.

7. The action plan is implemented (adapted from Checkland and Scholes 1999; Midgley 2000).

Directly applying these seven steps is known as 'mode one' of soft systems methodology. 'Mode two', in contrast, is the application of the general idea of the methodology—namely, comparing models of the future with participants' understanding of the current situation, without necessarily following the seven steps. The principles are internalised, as Checkland and Scholes (1999) put it, leaving the practitioner free to use any methods that seem appropriate. The application of mode two has led to some confusion as to just what is meant by soft systems methodology (Holwell 2000).

Soft systems methodology is a mature, well-tested dialogue method that has been applied in many ways in many settings. It is not, as some have claimed, a simple substitution of objectivity with subjectivity in systems thinking. Rather, as Midgley clarifies, 'the emphasis is on *inter*-subjectivity: the acceptance of multiple worldviews and the evolution of mutual understanding through debate' (Midgley 2003:vol. 1, p. xxxvii, emphasis in original).

Soft systems methodology is particularly useful when a need exists to develop realistic action plans to address complex (social) situations in which people are confused about, or hold differing views about, the nature and origins of the problem, how it can be addressed and what goals are to be worked towards. The group process helps people to attain a shared judgment that can be the basis for action to overcome the problematical situation.

Example of its use in research integration

Security: determining the most appropriate counselling model for community agencies' responses to disaster in Northern England

What was the context for the integration?

A key aspect of responding to large-scale disasters is synchronised activity between many different agencies, but the necessary collaborations are difficult to develop, maintain and implement. The human service agencies of a county in Northern England were concerned that, in the event of a disaster, they would not be able to engage in effective multi-agency activity to provide counselling services to the affected populations. A working party had been meeting for 18 months attempting, without success, to develop a multi-agency intervention plan. The barriers to achieving their goal were the complexity of the task and the differences of opinion as to the most appropriate model to use: one based on professional counselling services versus one based on volunteer counsellors' contributions (Gregory and Midgley 2000).

What was the integration aiming to achieve and who was intended to benefit?

Two consultants were invited to assist, and decided that soft systems methodology would be helpful as the need existed 'to structure the problems and facilitate debate', to develop models of inter-agency collaboration and to integrate the conflicting visions as to how collaboration could be realised.

What was being integrated?

Representatives from 19 agencies concerned about providing counselling services in the event of a disaster participated in this multi-agency activity, including health authorities, the ambulance service, the fire brigade, police, the Police Welfare Service, Victim Support, CRUSE (a voluntary organisation offering bereavement counselling), the Samaritans, a local Association of Counsellors, the Emergency Psychological Service, the Council of Churches, university departments, Emergency Planning (County Council) and Social Services (County Council).

Who did the integration and how was it undertaken?

As mentioned above, representatives from 19 agencies concerned about providing counselling services in the event of a disaster were involved, meeting in three blocks of two days each, during a one-month period. The consultant facilitators used soft systems methodology's mode one, following the seven steps with some small modifications. They began with an exercise to explore the nature of a disaster and then moved on to produce 'rich pictures'. At this stage in the process, participants were not positive about where it was leading, as they had brought to light many difficult and interrelated problems, with no solutions being apparent.

In the next step, participants were asked to identify the systems that would be needed to establish and implement the multi-agency counselling network if a disaster occurred. Since many were identified, they were asked to select the most important and to explore them in more detail using the CATWOE approach described above (although root definitions were not developed). The next step (a departure from the standard mode one approach) was to engage in whole-system modelling, after which detailed conceptual modelling was done of six of the systems identified. More conceptual modelling was not possible owing to time constraints. In this way, participants were assisted to reach accommodation of their differing visions for how the agencies could collaborate in the event of a disaster. An action plan was developed and used as the basis of an application for funding to establish the multi-agency network.

What was the outcome of the integration?

The evaluation of the process addressed participants' learning from the soft systems methodology process and the contents of the model that had been developed. Participants were positive about the process and indicated that they had learned a lot about the needs and priorities of the various agencies, and about the soft systems methodology itself. With respect to the contents of the model, although no disasters had occurred in the county within the two years after the development of the model, one did occur in a neighbouring county. Counselling support was provided to the people in that area in an effective and timely manner—a good test of the arrangements that had been developed through the soft systems methodology exercise.

Commentary

Soft systems methodology is a less structured dialogue method than many others. Especially in its 'mode two', it is best seen as a process, an approach and a perspective, as well as a method. The example illustrated this in its flexible application of the standard six CATWOE steps.

It focuses on action within a systems perspective. The purposeful action analysed is a system itself interacting with other systems.

A core issue for this method is developing participants' understanding of multiple world views and identifying how to accommodate them. This is in contrast, for example, with strategic assumption surfacing and testing, which tries to change world views or produce a new shared one among participants.

Also significant is the method's focus on developing agreed action plans and implementing them, rather than concluding the process with a set of integrated judgments (for example, through a consensus development panel's deliberations) where the dialogue itself (rather than follow-up action) is a valued outcome.

We have not been able to identify a case example that shows soft systems methodology being used for research integration but are confident that it has real potential for this. For example, the method could be used by a team of researchers and natural resource management planners wishing to develop an evidence-based action plan to deal with a complex issue such as the withdrawal of irrigation rights from farmers in a small rural community. The differing knowledge, world views and perspectives held by these actors would need to be integrated to produce shared judgments of the likely impacts and trade-offs inherent in any action plan that they would develop.

Origins and genealogy

Checkland and, later, other systems scholars and practitioners developed soft systems methodology as a response to the limitations that they saw in the reductionist approaches of the natural sciences when these were applied to complex social situations. The dominant systems thinking approach at the time (the 1970s) was the 'hard' paradigm of systems engineering: defining the system of concern, defining the system's objectives, then engineering the system to meet those objectives (Checkland 1985). Soft systems methodology, in contrast, was developed 'because the methodology of systems engineering, based on defining goals or objectives, simply did not work when applied to messy, ill-structured, real-world problems. The inability to define objectives, or to decide whose were most important, was usually part of the problem' (Checkland 1985:763). Soft systems methodology was designed to overcome these limitations.

Further reading on soft systems methodology

Checkland, P. 1981, *Systems Thinking, Systems Practice*, J. Wiley, Chichester, Sussex.

Checkland, P. and Poulter, J. 2006, *Learning for Action: A short definitive account of soft systems methodology and its use for practitioners, teachers, and students*, John Wiley and Sons, Hoboken, NJ.

Checkland, P. and Scholes, J. 1999, *Soft Systems Methodology in Action*, Wiley, New York.

Midgley, G. 2000, *Systemic Intervention: Philosophy, methodology, and practice,* Kluwer Academic/Plenum Publishers, New York.

Endnotes

[1] It should be noted that other organisations also conduct consensus development conferences, many using that term and using various processes with varying degrees of similarity to the US National Institutes of Health's approach discussed here. For example, in 2005, a Consensus Conference on Cochlear Implant Soft Failures was held as part of the Tenth Symposium on Cochlear Implantation in Children in Dallas, Texas. Its consensus statement has been published in a medical journal (Balkany et al. 2005).

Chapter 4. Dialogue methods for understanding particular aspects of a problem: integrating visions, world views, interests and values

Introduction

Having discussed 10 dialogue methods for integrating judgments, we turn now to four methods that are useful for understanding particular aspects of a problem—namely, visions, world views, interests and values. A dialogue method addressing each of these is described, illustrated and discussed in this chapter. The methods are: for integrating visions — appreciative inquiry; for integrating world views — strategic assumption surfacing and testing; for integrating interests — principled negotiation; and, for integrating values — ethical matrix.

Integrating visions

We use vision here in the sense of a mental view or image of a goal that does not yet exist in place or time. Visions are important in research in terms of the overarching aspirations that a particular study seeks to contribute to. Integration is important for developing a shared vision or for accommodating different visions. For example, as we discussed earlier, within the same study, some researchers could have a grand vision such as alleviating national poverty, while others could be focused on improving employment opportunities for a particular group.

As with all the specific methods, we suggest that a method for integrating visions is used only when this is particularly salient in the research integration process. For many research questions, the issue of accommodating different visions will not be particularly important and a broad integration method, such as those outlined in the previous chapter, will be more than adequate. Sometimes, however, ensuring an understanding of the different visions of the research participants is essential to moving forward on integration and in such cases the method described here can be particularly useful.

Dialogue methods focusing on visions can also be helpful in setting an overarching vision, especially when this might motivate and direct the activities of the research team. For example, rather than the research just being conducted for its own sake, there might be a higher goal that the research can contribute to. Research on improving integrity systems in policing, for example, can be construed in terms of improving policing operations narrowly or as enhancing the role of policing in contributing to the rule of law in a democracy.

The research integrator would be expected to identify that integrating visions was salient to addressing a particular research question. Sometimes this would be evident before the dialogue process started; at other times, it would become clear once the process was under way. They would then take the lead in organising the dialogue to integrate visions. The purpose is generally to align the overarching goals of the members of the research team and the stakeholders to help them work to a common outcome, or at least compatible goals, in order to smooth the path of the research and its implementation.

We found one method for integrating visions: appreciative inquiry.

Integrating world views

World views or mental models are the underlying assumptions about how the world works that guide our understanding and actions. In a research integration context, the world views of those involved in the research are likely to include assumptions about the problem being addressed and the research process.

In terms of the problem being addressed, there could be differing assumptions about the importance of various aspects of the problem and the roles of diverse actors. For example, in a research project about heroin overdoses, there could be different assumptions about the role that drug-using peers can play. Some of those involved in the research might assume that peers were generally present at an overdose and would take action if they knew what to do. Others might assume that most overdoses occurred when someone used alone and, in any case, even if someone else was present, they would be more likely to run away than to render assistance. These differences in assumptions play out in terms of the stakeholders included in the dialogue and the questions asked of them. In the example presented, one group would argue for heroin users to be included in the dialogue and would ask about their actions in the case of an overdose, whereas the other would see these stakeholders and that question as irrelevant.

In terms of the research process, there could be differing assumptions about the purposes of the dialogue, the importance of various aspects of the process, whether or not consensus should be reached and what should be done with the results. For example, some participants might assume that everyone involved had an equal standing, that reaching consensus was crucial and that the relevant decision makers would act on the results. Others might assume that the judgments of more powerful groups had greater salience, that there just needed to be a show of consensus to end the process and that decision makers would be presented with the results to consider along with other inputs. If such differences in assumptions were not made explicit and were not resolved, participants might find themselves at loggerheads, without really understanding why.

Identifying differences in assumptions is often a positive rather than a negative experience. It can be energising to realise that others view the world differently and it can open up a broader range of possibilities for understanding and action. For example, if one group of participants in a dialogue had the world view that young people were responsible for their own alcohol consumption and another assumed that drinking was greatly influenced by availability, advertising and the actions of licensees, bringing them together could provide a richer understanding and a multi-pronged action strategy.

As with the other specific aspects of research integration, the research integrator would be expected to identify that integrating assumptions was salient to addressing a particular research question, either before or during the process. They would take the lead in organising the dialogue to integrate assumptions. This could be used to enrich the conversation or to remove stumbling blocks.

We found one method for integrating world views: strategic assumption surfacing and testing.

Integrating interests

Interests are what motivate us. Making a profit, personal advancement, concern about those less fortunate and a desire to protect a piece of wilderness are all examples of interests. Such motivations provide the reason why stakeholders and researchers choose to tackle a particular problem. As well, there are interests for becoming involved in a particular research project. Individual researchers could be motivated by the chance of publication, access to a specific data set, the opportunity to work with a particular individual and so on. For community groups, decision makers and other stakeholders, interests in becoming involved in the research could include wanting to see a problem gain legitimacy, ensuring that their point of view is heard or wanting to see something done about a problem.

Interests are important for research integration because conflicting interests can prevent progress from being made on an issue. Resolving such clashes in motivations can be essential for research to lead to effective decisions and practice-based change.

Negotiation is the usual method for resolving divergent interests, but many forms of negotiation are about one side winning at the expense of the other. These are not consistent with the aim of dialogue to 'jointly create meaning and shared understanding' (Franco 2006:814). One form of negotiation, however—principled negotiation—stands out as different and is consistent with the aims of dialogue.

Principled negotiation is the only form of dialogue for integrating interests that we have come across to date and is therefore the only method dealt with here. While it was developed as a conflict-resolution tool, integrating interests using principled negotiation can occur before conflict comes to a head. For example, it can be useful early in research when an understanding of different motivations can be used to shape the details of the research and the rewards for the various research participants.

The research integrator would be expected to identify, as with the other specific aspects of research integration, that integrating interests was salient to addressing a particular research question, either before or during the process. They would take the lead in organising the dialogue to integrate interests.

Integrating values

Values are the moral stance that underpins the research. We use a definition from *The Oxford English Dictionary* (1989): 'the principles or standards of a person or society, the personal or societal judgement of what is valuable and important in life.'

Values are important in research integration because, as with visions, world views and interests, appreciating differences in values can enrich the understanding of the research question, as well as identifying potential or real sources of conflict.

The importance of accommodating different values in research of real-world problems is becoming increasingly recognised. In the area of natural resource management research and practice, for example, Lockwood (2005) described methods for including values in environmental choices. Further, a project on ethical tools funded by the European Union (<http://www.ethicaltools.info/>) set out to improve ethical assessment 'by broadening the values considered and/or stakeholder involvement'. The researchers in that project were particularly concerned about the ethical issues involved in the introduction and application of new technologies in agricultural and food production and pointed out that it was unlikely that a single ethical tool would be adequate for a full ethical assessment—a point also made by Lockwood (2005). One of the tools to which they point is the ethical Delphi, an application of the Delphi technique to ethical issues (Millar et al. 2007).

The research integrator would be expected to identify, as with the other specific aspects of research integration, that integrating values was salient to addressing a particular research question, either before or during the process. They would take the lead in organising the dialogue to integrate values. This could be used to enrich the conversation or to remove stumbling blocks.

We deal here with one of the methods the European Union researchers recommended: the ethical matrix. They also pointed out that the consensus conference and the Delphi technique could be modified to deal with values.

Appreciative inquiry: integrating visions

Description

Appreciative inquiry is a dialogue method that brings together members of an organisation to clarify, develop and integrate their visions about their joint endeavours by identifying what is good about their visions and how to move the organisation to a higher level of goal attainment. The process produces better understanding of the present situation and future possibilities and facilitates better decision making through its focus on the team's strengths. While it is generally used in an organisational context, it appears that it can be readily applied to research integration.

Appreciative inquiry is based explicitly on a constructivist world view—that is, an understanding that our perceptions of reality are socially produced and reproduced, rather than representing fixed, external realities. An important implication of this position for appreciative inquiry is that our realities are open to change when we see the world through different eyes.

Appreciative inquiry works from a set of eight assumptions.

1. In every society, organisation or group, something works.
2. What we focus on becomes our reality.
3. Reality is created in the moment, and there are multiple realities.
4. The act of asking questions of an organisation or group influences the group in some way.
5. People have more confidence and comfort to journey to the future (the unknown) when they carry forward parts of the past (the known).
6. If we carry forward parts of the past, they should be what is best about the past.
7. It is important to value differences.
8. The language we use creates our reality. (Hammond 1998)

The most important characteristic of appreciative inquiry is its focus on what has worked well within a team or organisation, in contrast to the more common approach of identifying problems (what has not worked well) and analysing these. As appreciative inquiry proponents regularly point out, if 95 per cent of a company's clients express satisfaction with how the company meets their needs, why bother to analyse the 5 per cent of failures when far more can be learned, and the company can become far more effective, by analysing what is going well with the other 95 per cent?

The implementation of appreciative inquiry occurs in four phases:

- appreciating and valuing the best of 'what is'
- envisioning what 'might be'
- dialoguing 'what should be'
- envisioning 'what will be'.

This covers the '4-D' cycle of appreciative inquiry: discovery, dream, design and destiny.

People who use appreciative inquiry in organisational development and other applications report that participants in the group discussions readily become energised, animated and captured by the power of the process, with its emphasis on what works rather than on deficits, and the shift in shared attitudes from fixing problems to building on past successes.

In terms of research integration, appreciative inquiry is a method to clarify visions of the researchers and stakeholders and to draw them together into a shared vision for the future of the research. That then forms the basis of decision making about how the research team might operate in the future, and what goals it should work towards attaining.

Examples of its use in research integration

1. Public health: meeting the needs of elderly people in transition from hospital to the community

What was the context for the integration?

Significant challenges exist in the health and social welfare systems in meeting the needs of elderly people discharged from hospital into the community, many without adequate support available from their families or community agencies.

What was the integration aiming to achieve and who was intended to benefit?

Appreciative inquiry was used to bring together people from 37 agencies who were concerned about this problem (Reed et al. 2002). It involved clarifying agency personnel visions for better hospital discharge experiences for older people and developing action plans to help turn these visions into reality. The process was structured explicitly to elicit a variety of different visions and to narrow these down to a small number, agreed on by participants, that could be turned into realistic action plans.

What was being integrated?

The agencies invited to participate included older people's organisations, community and hospital trusts, local government, community organisations and others in the healthcare sector. In addition to these key groups, other significant organisations, such as gas and electricity suppliers, participated. The individuals involved had a variety of backgrounds, including health professionals such as nurses and doctors, and lay people from older people's organisations.

Who did the integration and how was it undertaken?

The appreciative inquiry exercise was initiated by sending letters of invitation to 37 local agencies in the Newcastle Health Authority in the United Kingdom. Seventy-one people expressed interest in participating.

Three workshops were held. The first introduced the appreciative inquiry method and trained participants in it. In particular, they developed interviewing and recording skills. They were tasked to interview people in their organisations, the community and so on using the core appreciative inquiry dialogic approach. This entailed the following steps:

- Interviewers asked people to tell them about a time when they felt that a hospital discharge had gone well
- Interviewers then asked probing questions to explore what people had valued about their contribution to that discharge experience and what they felt had helped them to contribute
- Interviewees were asked to imagine that a miracle had occurred overnight that had helped discharges to go well every time
- They were then asked to tell the interviewer what would be different about the world after the miracle, what would be in place, what would be happening and what the results would be (Reed et al. 2002).

At the second workshop, the data from the interviews were discussed in detail to identify what the participants thought were the key themes. As part of this process, the facilitators combined a nominal group approach with appreciative inquiry to ensure that all had an equal opportunity to contribute to the evolving consensus. At the third workshop, provocative propositions were developed and discussed, and from that action plans were developed.

This process brought to light the diverse visions for better post-hospital experiences for older people in the community and collapsed and prioritised these into a manageable set of themes and action plans. The breadth of the visions elicited and integrated through the process is revealed in the following list of provocative propositions developed in the third workshop:

1. Every worker/patient/carer knows exactly what other workers do
2. Every worker has the opportunity within their job to develop and use networks across the system
3. Every worker takes responsibility to act on information that they hold or receive about the person and their circumstances

4. There is a process for checking up to ensure that the person going home is managing and has received all the services they were expecting
5. There is co-ordination of the process of going home to avoid duplication and/or gaps in the services and support provided
6. Resources and funding are co-ordinated to avoid disputes or time-consuming negotiations
7. All patients are given individually tailored information and support to ensure choice and ownership of their own health/social care
8. There are joint care plans (held by the patient) and shared documentation/information
9. Care is flexible and responsive and there is a fall back strategy. (Reed et al. 2002:41)

What was the outcome of the integration?

The researchers responsible for this process (Reed et al. 2002) reported that most of the participants were pleased with the appreciative inquiry process to which they had contributed. They pointed out, though, some of its limitations, including the perception that it ignored problems (by focusing on the positives), the challenges to the evaluation of the process and the fact that (on this occasion) the quality of the action plans was adversely affected owing to the limited participation in the process by high-level decision makers. Nonetheless, this is an example of the integration of the visions of a range of actors keen to improve the organisational responses to older people's needs in such a manner as to produce a set of action plans to make that a reality.

Action plans were developed based on the propositions developed. These were as follows:

- the development of an information pack, to be accessible to all patients, carers and agencies, about support services and resources for people going home from hospital
- holding an event for people involved in training and staff development
- each agency should identify key people to oversee the process, including follow-up
- a person-centred care plan should be developed that is easy to understand and that includes the views of users and carers. The group knew of some developments in this area and agreed to explore them further (adapted from Reed et al. 2002).

2. Organisational development: developing a code of ethics in a university department in South Africa

What was the integration aiming to achieve and who was intended to benefit?

The appreciative inquiry approach was used in a department within a South African university to develop a code of ethics and an approach to ethical management that realised a shared vision of 'what "could be", rather than "what was"' with respect to departmental ethics (van Vuuren and Crous 2005:403).

Who did the integration and how was it undertaken?

Thirty-six members of a university department attended a 12-hour appreciative inquiry intervention, spread over three sessions. Group processes facilitated by two industrial psychologists (the processes were not described) were used to clarify the topic ('a vision for ethics'), the discovery phase identified what was going well and desires for an ethical future, and the dream phase identified 18 important issues that were distilled to four core themes. In the design phase, provocative propositions were developed and discussed and the destiny phase produced the new code of ethics and plans for its implementation.

What was the outcome of the integration?

Participants were unanimous that the process and product exceeded their expectations. Through the appreciative inquiry approach to dialogue, participants of all ranks in the organisation were able to develop and share their individual visions for its ethical future and produce an integrated vision, which all shared, in the form of a code of ethics and an action plan for its implementation. The appreciative inquiry method created better understanding of the issues and the way forward, integrating the visions of the members of the department about how ethics could become core business within the organisation.

What was being integrated?

All the academic and administrative staff of the university department were invited to participate and 36—all but two—agreed to do so. Further information about the participants was not provided.

What was the context for the integration?

The appreciative inquiry exercise was conducted within an academic department of a large, state-funded South African university: the University of Johannesburg. It was undertaken in the context of improving the governance of the department, specifically to realise the staff's desire to create 'an ethical way forward' for the department (van Vuuren and Crous 2005:408).

Commentary

The two case studies show that appreciative inquiry focuses on developing a shared vision, rather than trying to accommodate different visions. The value of the process is that it leaps ahead with the assumption that a shared vision is important and achievable, rather than trying to clarify and accommodate differences in current visions. This will not be beneficial for all cases of research integration, so other methods for integrating visions will also be required.

Both of the cases deal with stakeholders only and disciplinary knowledge is not brought to bear. Even though the second case involves members of an academic department, disciplinary expertise is not relevant to moving forward on the problem in this instance.

In both cases, researchers organised the dialogue, but there was little description of their roles. Both cases were focused on improving practice rather than improving understanding about an issue.

The embedding of the nominal group technique in the first case is particularly interesting and illustrates the diverse ways in which dialogue methods can be used. For simplicity, in our general presentation, we have characterised specific methods, such as appreciative inquiry, as subservient to or embedded in more general methods, such as the nominal group technique. In this case, however, appreciative inquiry and the question it addresses are clearly dominant, with the nominal group technique playing a subservient role in one part of the appreciative inquiry process.

While research integration, as we define it in this book, does not seem to be the usual way in which appreciative inquiry is used, it is possible to imagine its use by a research team, comprising disciplinary experts and stakeholders, to develop an overarching goal for a particular project. For example, one can imagine a team, such as the commissioners of the World Commission on Dams, using a method such as appreciative inquiry to develop the overarching goal of the commission to achieve 'development effectiveness', where 'decision-making on water and energy management will align itself with the emerging global commitment to sustainable human development and on the equitable distribution of costs and benefits' (see also Bammer 2008a; World Commission on Dams 2000:xxxiii). We are not suggesting that this is what the World Commission on Dams has really done, but it provides an example of research integration that readers might be able to identify with.

Origins and genealogy

David L. Cooperrider coined the term 'appreciative inquiry' from his doctoral research at Case Western Reserve University, Cleveland, Ohio, in the first half of the 1980s. The ideas and methods he developed were rapidly taken up and extended by others. Appreciative inquiry is still developing and being applied in many different sectors under the leadership of Cooperrider and others.

Further reading on appreciative inquiry

Appreciative Inquiry Commons n.d., *Appreciative Inquiry Commons*, Weatherhead School of Management at Case Western Reserve University, <http://appreciativeinquiry.case.edu/>

Cooperrider, D. L. and Whitney, D. K. 2005, *Appreciative Inquiry: A positive revolution in change*, Berrett-Koehler, San Francisco, Calif.

Hammond, S. A. 1998, *The Thin Book of Appreciative Inquiry*, Second edition, Thin Book Publishing Company, Plano, Tex.

Hammond, S. A. and Royal, C. (eds) 1998, *Lessons From the Field: Applying appreciative inquiry*, Practical Press, Plano, Tex.

Watkins, J. M. and Mohr, B. J. 2001, *Appreciative Inquiry: Change at the speed of imagination*, Practicing Organization Development Series, Jossey-Bass/Pfeiffer, San Francisco, Calif.

Strategic assumption surfacing and testing: integrating world views

Description

Strategic assumption surfacing and testing is a method for integrating world views developed in an organisational context. It is based on the premise that '[w]e all live our lives according to the *assumptions* we make about ourselves and our world. To cope better, we need to surface those assumptions and to challenge them. New assumptions then become springboards to effective change' (Mason and Mitroff 1981:vii, emphasis in original).

This method assists participants to understand a problematic situation and explore strategies for dealing with it. As the name indicates, its central element is bringing to the surface the assumptions that underlie people's preferred approaches to an issue, and challenging them. Sometimes this challenging results in a particular strategy being discarded and participants adopting a competing one. On other occasions, however, integration occurs through the synthesis of previously inconsistent assumptions, resulting in a new strategy that accommodates the differences between those held initially, and which is stronger than the components from which it arises.

Four principles underlie the strategic assumption surfacing and testing method: it is adversarial, participative, integrative and 'managerial mind supporting'.

- *Adversarial*—based on the belief that judgments about ill-structured problems are best made after consideration of opposing perspectives.
- *Participative*—it seeks to involve different groupings and levels in an organisation, because the knowledge and resources needed to solve complex problems and implement solutions will be distributed around a number of individuals and groups in the organisation.
- *Integrative*—on the assumption that the differences thrown up by the adversarial and participative processes must eventually be brought together again in a higher order synthesis, so that an action plan can be produced.
- *Managerial mind supporting*—believing that managers exposed to different assumptions will possess a deeper understanding of an organisation, its policies and 'problems'. (Flood and Jackson 1991:123–4)

The four steps used in the method are as follows:

- *Group formation:* gathering as many as possible of those involved in, and affected by, a situation and splitting them into small groups according to their views on key issues. It is important to minimise the conflicts *within* each group and to maximise the differences *between* groups. The orientation to the problem held by each group should be directly opposed by at least one other group.

- *Assumption surfacing and rating:* identifying the preferred strategy or position that each group is adopting, then revealing and quantifying (if possible) the assumptions on which it is based. Techniques used include stakeholder analysis, assumption specification and assumption rating.
- *Intra-group and inter-group dialectical debate:* each group developing the case for its position and then discussing them all in a single, large group. The process is dialectical 'if it examines a situation completely and logically from two different points of view' (Mason and Mitroff 1981:129). A key analytical question that facilitates dialogue in this stage is 'What assumptions of the other groups do each group find the most troubling?'.
- *Final synthesis:* achieving an accommodation among participants to find a practical way forward. Discussion of key assumptions leads them to be modified and a new strategy to be developed, based on the modified and agreed-on assumptions. This is the process through which the visions and world views of the participants become integrated. If agreement cannot be reached—if synthesis is not possible—participants might agree on a program of research or other action to further clarify assumptions and/or to try out a particular strategy and evaluate it. Knowledge gained from those steps can shed further light on the conflicting assumptions, facilitating subsequent synthesis of positions (adapted from Flood and Jackson 1991; Mason and Mitroff 1981; Midgley 2000).

This approach to strategic planning has been contrasted, by its originators, with the two dominant approaches—namely, the 'expert' approach, in which an organisation establishes a planning unit to largely do the managers' work for them, and the 'devil's advocate' approach, in which middle managers prepare and submit plans to senior managers for cross-examination (Mason and Mitroff 1981:127–9).

The strategic assumption surfacing and testing method was developed as a contribution to strategic planning. It has great potential where conflicting views on the nature of a problem and what to do about it are held, where the proponents are willing to work in groups to explore these issues and are open to hearing and understanding others' views, with the aim of finding accommodation between the originally conflicting positions. Being willing to reveal, explore and expose to criticism the assumptions that one brings to the process, and a concomitant willingness to challenge others' assumptions, are essential to the achievement of synthesis.

Example of its use in integration

Despite the potential of the strategic assumption surfacing and testing method for research integration, we have not been able to find any recent cases that illustrate this well with respect to the specific role of research. As a result, the following case example comes from the business sector.

Business: seeking agreement on the core operational strategy of a Cooperative Development Agency in the United States

What was the context for the integration?

A Cooperative Development Agency in the United States—known as 'Winterton' for the purposes of the case study—had the aim of fostering and promoting commercial and industrial activity in its county. As with all Cooperative Development Agencies, it worked to achieve its goal through cooperative enterprises—that is, business entities owned and usually managed by the people who worked in them (Flood and Jackson 1991).

What was the integration aiming to achieve and who was intended to benefit?

Agency staff wished to analyse the agency's methods of operating, improve its marketing activities and identify ways to more efficiently serve the people in the county in which it was located. They wished to identify the optimal organisational structure for attaining its goals. Their overarching goal was to improve the quality of services to the people of the county while remaining true to the values and norms of the cooperative movement.

What was being integrated?

The staff of the agency worked together with the aim of developing agreement about how the organisation should be structured so as to best implement the shared values of the cooperative movement. Different staff members had strongly conflicting visions about the optimal organisational arrangements, and different stakes in the outcomes, which needed to be synthesised for the organisation to achieve its goals within its business environment.

How was the integration undertaken and who did the integration?

The staff of the Cooperative Development Agency, with the support of external expert facilitators, attempted to integrate their world views. They used the soft systems methodology to do this. At an early point, it became clear that participants in the process fell into two opposing factions, one of which favoured a top-down approach to the agency's operations and the other a bottom-up approach. The top-down approach was one in which the agency identified business opportunities and recruited people into cooperatives to respond to those opportunities. The bottom-up approach, more closely reflecting the norms of cooperatives, emphasised assisting people thinking about engaging in business to form cooperatives and then seek out business opportunities.

Strategic assumption surfacing and testing was used to deal with this deep conflict; the aim was to do so quickly, without getting into the details of the governance arrangements but simply to reach agreement—a synthesis of assumptions—so that the soft systems methodology exercise could continue. In this sense, the strategic assumption surfacing and testing method was nested within the dominant soft systems method.

The standard four stages of the method were followed. *Group formation* presented no challenges, with the staff readily falling into two groups: one strongly favouring the top-down approach and the other the bottom-up approach to the operation of the agency. The two groups were separated and each went through the *assumption surfacing* stage by using stakeholder analysis, assumption specification and assumption rating. The two groups identified different groups of stakeholders and the assumptions linked to each. For example, the top-down supporters identified the unemployed as stakeholders, along with an assumption about the credibility of the agency, in the eyes of funding bodies, with respect to job creation. The bottom-up group identified potential clients as stakeholders along with the assumption that they lacked group cohesion. The two groups joined together again for *dialectical debate*. They found little common ground and little ability to modify their differing lists of key stakeholders and the assumptions linked to each, with the result that no *synthesis* emerged from the dialectical debate.

What was the outcome of the integration?

Although the process did not produce synthesis of the assumptions of participants, it nonetheless had some positive outcomes: 'Consensus was…reached on particular matters such as the need to seek out sources of information about business opportunities, to research other top-down experiences, and on the desirability of some experiments with a modified top-down approach (which were, indeed, carried out)' (Flood and Jackson 1991:132).

Commentary

This example embeds strategic assumption surfacing and testing within a broader soft systems methodology. It is an excellent demonstration of the point we make more generally about how a specific method can be used to resolve a challenge discovered when a method for achieving broader understanding is used. In this case, however, it was not successful in resolving the conflict that became evident.

It is noteworthy that we could not find an example of the application of this method in a research integration context. It is, however, conceivable how it could be used in this way. If we consider, for example, a research question about how the health sector should respond to violent clients, we could imagine bringing together various stakeholders and exposing their assumptions about clients' responsibility for violence. We might also add disciplinary perspectives—for example, from psychology, sociology and clinical research.

This dialogue method is also noteworthy for being intentionally adversarial: dialectic debate about the validity of people's assumptions is at its core.

Origins and genealogy

This method is part of what Midgley (2000:193) refers to as the second wave of systems thinking in which '"systems" were no longer seen as real world entities, but as constructs to aid understanding. The emphasis was on dialogue, mutual appreciation and the inter-subjective construction of realities.' Mason and Mitroff (1981) adapted the approach to systems thinking explicated by C. West Churchman, turning some of his ideas into the step-by-step strategic assumption surfacing and testing method.

Further reading on strategic assumption surfacing and testing

Dash, D. 2007, *SAST Methodology*, Xavier Institute of Management, Bhubaneswar, India,
<http://www1.ximb.ac.in/users/fac/dpdash/dpdash.nsf/pages/CP_SAST>

Flood, R. L. and Jackson, M. C. 1991, *Creative Problem Solving: Total systems intervention*, Wiley, Chichester, New York.

Mason, R. O. and Mitroff, I. I. 1981, *Challenging Strategic Planning Assumptions: Theory, cases, and techniques*, Wiley, New York.

Midgley, G. 2000, *Systemic Intervention: Philosophy, methodology, and practice*, Kluwer Academic/Plenum Publishers, New York.

Principled negotiation: integrating interests

Description

An important challenge for research integration is to accommodate the different interests of the researchers and other stakeholders. Principled negotiation, also known as negotiation on the merits or 'getting to yes', is an effective way of achieving this.

Fisher et al. (1991) have developed four steps for principled negotiation:

1. separate the people from the problem
2. focus on interests, not positions
3. generate a variety of possibilities before deciding what to do
4. insist that the results are based on some objective standard.

Separating the people from the problem requires recognising that any problem has two components—the relationship and the substantive or content issue—and that they often become intertwined. In terms of problem solving, there are three aspects of relationships to be mindful of: emotions, perceptions and communication.

With respect to *emotions*, a key element of principled negotiation is to recognise and understand our own emotions and those of our partners. Sometimes it can be helpful to make explicit how we are feeling and to encourage our collaborators to do the same, thus acknowledging the legitimacy of emotions. The aim of principled negotiation is to channel emotions into a productive vision of working side by side to bridge differences (or, if there is conflict, to attack the problem).

With respect to *perceptions*, everyone has their own version of reality. A common problem is that we misinterpret others' intentions. It is important to try to see the issue from the other person's perspective, while acknowledging that understanding is not the same as agreeing.

With respect to *communication*, clarity in this area is important for bridging differences or solving mutual problems. Essentially, communication founders when people do not listen, do not hear, misunderstand or misinterpret. In brief, it is important to listen actively in order to better understand, and to be tolerant and slow to take offence.

Focusing on interests, not positions recognises that it is generally hard to find mutually satisfactory resolution between competing positions. It is therefore important to shift the focus from positions to interests.

Behind opposed positions lie shared and compatible interests, as well as conflicting ones. The process of identifying interests therefore usually clarifies where real disagreements lie and, because some interests will be shared or complementary, the areas for conflict will generally be smaller than first thought.

Listening with respect, showing courtesy and emphasising concern to meet the basic needs of the other person direct the focus to interests. In addition, people applying this method try to be specific about their own interests and how important they are to them. The goal is to frame a joint attack on the difference or problem that will accommodate the interests of both parties.

Generating a variety of possibilities before deciding what to do. When using the *principled negotiation method*, generating options for action can often be the hardest step to get partners to participate in. Fisher et al. (1991) suggest that there are four particular obstacles to this: premature judgment, searching for a single answer, assumption of a 'fixed pie', and thinking that no-one else is able to assist in problem solving.

The temptation to leap to a solution before considering the options is often coupled with the assumption that there is only one 'right' answer, rather than an appreciation that there are generally many ways in which interests can be met. Further, it can also be a trap to assume that there are no additional resources that can be brought into play (a 'fixed pie'). The key is to see the areas of difference or conflict as *shared* problems requiring *shared* solutions. Seeing them as simply the partner's problem can result in the partner developing solutions that do not take our interests into account.

The most widely used method for generating a variety of possibilities is brainstorming, in which participants are encouraged to rapidly put forward ideas, while at the same time withholding judgments on their merits. Encouraging interaction at speed, without in-depth discussion, tends to circumvent narrow thinking and opens up the possibility of creative solutions. The idea is to search for mutual gains, to dovetail different interests and, if necessary, to give all partners an easy way of backing away from previously stated positions. It is essential to look forward and to leave past disagreements to one side.

Looking for a fair solution, based on the merits. Once options have been generated, the next step is to evaluate them and to find a fair solution based on their merits. It helps to be concrete but flexible—in other words, to work through the options in detail, but to treat the options as illustrative rather than fixed. The commitment has to be to address participants' interests (not positions) and, by pushing hard on the interests, partners can stimulate each other's creativity in thinking up mutually advantageous solutions. 'Be hard on the problem, soft on the people.' The aim here is to move the solution away from a notion of partners giving in to each other towards one in which both are deferring to a fair solution. They are yielding to principle, not pressure. Developing an agreement should be framed as a joint search for objective or fair criteria and this will promote reasonableness, fair play and trustworthiness.

Principled negotiation is a powerful dialogue method in research integration where strengthening partnerships through addressing conflicting or apparently conflicting interests is important. It promotes accommodation of different interests not only between researchers, but between researchers and policy-makers, service providers, businesses and community groups. In a recent follow-up to the original work on principled negotiation, Fisher and Shapiro (2005) identify five 'core concerns' that motivate people: appreciation, affiliation, autonomy, status and role. Considering these can be useful in teasing out the dimensions for accommodation in a research project. When disputes in integrative research arise, this method can also be useful for resolving them.

Example of its use in a public health problem: a social worker uses principled negotiation to advocate for their clients' rights[1]

In the social work profession, principled negotiation has been characterised as 'a new tool for case advocacy', with 'case advocacy' being defined as a way to 'obtain resources or services for clients that would not otherwise be provided' (Lens 2004:506). For the social worker, advocating for clients' entitlements often entails discussions with officers of public sector agencies who have the power to provide or withhold helping services. Sometimes these discussions are confrontational, with the social worker demanding that the services be provided to their client and the official stating that they will not be provided because the client falls outside the entitlement rules, because of funding/service shortages, and so on. Principled negotiation is a viable—indeed, preferred—alternative to demanding that the services be provided, an alternative that integrates the interests of the public sector service gatekeeper with those of the social worker advocating for her or his clients.

This was illustrated in the case of a 95-year-old woman, living in her own home and suffering from a variety of debilitating medical conditions. She needed assistance with almost all her daily requirements and also needed overnight supervision as she was prone to wandering. The local government social service agency had been providing 24-hour, split-shift cover by personal carer aides but wanted to reduce this to a full-time live-in aide. The problems were that the aide needed to be awake and able to assist the client throughout the night and there was nowhere in the home for the aide to sleep at other times. The client's social worker used principled negotiation to challenge the department's decision to reduce the level of services and to advocate, on the client's behalf, with the officer responsible for the decision (Lens 2004). (It is not completely clear, in Lens' paper, if this is a report on a real case or a composite, illustrative example.)

The social worker applied four rules of principled negotiation: separating the person from the problem, focusing on interests not positions, inventing options for mutual gain, and insisting on using objective criteria.

- Separating the person from the problem was illustrated by the social worker refraining from complaining that the officer's decision reflected incompetence and insensitivity to the elderly. Instead, the point was made that '[m]y client is very ill and in need of continued services. How do you think we can help her?' (Lens 2004:508).

- Focusing on interests, not positions entailed maintaining a direct statement of the client's interests but not adopting a rigid position from the beginning of the principled negotiation. The officer stated that 'I must tell you again the agency does not have unlimited funds', to which the social worker responded, 'As I understand it, your interest as a government agency is to be fiscally cautious while also helping people...Perhaps there is a way we could work together to get the agency to provide appropriate coverage...it would avoid placement in a nursing home, which would cost the agency even more' (Lens 2004:510).

- Inventing options for mutual gain occurred in the exchange just quoted, in which the phrase 'appropriate coverage' was used rather than the social worker demanding a particular type of coverage. This was based on a brainstorming session undertaken to identify other options, before the negotiations were initiated.

- Insisting on using objective criteria was important in this case, applying the principled negotiation dictum that 'principle, not pressure' is a key to attaining a win-win position. The social worker had checked the details of the client's legal entitlements before initiating the negotiation. In discussion with the officer, the social worker refrained from exclaiming 'You don't know what you are doing and are unaware of the law', instead calmly stating, 'Please correct me if I am wrong, but it is my understanding that the regulations require a split shift when patients such as [the client] cannot do daily activities and their safety is jeopardized' (Lens 2004:511).

Throughout the principled negotiation session, the social worker maintained the focus on advocating for the client, but did so in a manner that revealed and dealt with the interests of both parties (the client and the government agency), acknowledging that both had valid interests that needed to be taken into account. It was found to be possible to integrate the interests of both parties to reach a mutually satisfactory, win-win outcome.

Commentary

Individuals and organisations have various interests and sometimes they conflict, or have the potential to conflict. It is not difficult to imagine circumstances in which members of a research team have interests that conflict (for example, several members of a team may want to be the first author on a publication). Furthermore, at the beginning of an interdisciplinary research project it would be possible to identify potential conflicts through principled negotiation and to deal with them before they become problematic. Since different interests inevitably exist in the interfaces between researchers and users of research in policy and practice settings, scope also exists here for principled negotiation to be used to avoid problems and/or deal with them as they arise.

We have not been able to locate any published reports of the use of principled negotiation by researchers seeking to improve research integration, so we have illustrated this method with an example from public health that does not have a research focus. Nonetheless, we see great potential for its application to research.

Origins and genealogy

Principled negotiation, as described here and documented by Fisher et al. (1991), is a product of the Program on Negotiation based at the Harvard Law School (<http://www.pon.harvard.edu/>). The program was established in 1983 and operates through a consortium of three universities: Harvard, Tufts and Massachusetts Institute of Technology (MIT). It builds on proposals concerning negotiation styles proffered in the early years of the twentieth century by Mary Parker Follett (the originator of the term 'conflict resolution') and others.

Further reading on principled negotiation

Fisher, R., Ury, W. and Patton, B. 1991, *Getting to Yes: Negotiating an agreement without giving in*, Second edition, Random House, London (a five-page summary is on the web site of the University of Colorado's Conflict Research Consortium, <http://www.colorado.edu/conflict/peace/example/fish7513.htm>).

Fisher, R. and Shapiro, D. 2005, *Beyond Reason: Using emotions as you negotiate*, Viking, New York.

Ethical matrix: integrating values

Description

The ethical matrix is a method developed by Mepham (2000) for rational ethical analysis. It comes from the discipline of applied ethics and is based on the acknowledgment that, in modern pluralistic society, the various actors involved in a given issue hold different and potentially conflicting values. Since decisions have to be made, and a potential exists for overriding the values of some stakeholders in the process, it could be useful to have available a structured process for surfacing, weighing and integrating those values in the decision-making process. The ethical matrix was developed to meet those needs.

The ethical components of the method reflect 'commonsense morality'. They are derived from the work of Beauchamp and Childress (2001) on bioethics that has been widely taken up in medicine and medical ethics. Beauchamp and Childress introduced the 'four principles approach' through which decision makers were guided to consider four core values: non-maleficence (doing no harm), beneficence, autonomy and justice. Mepham has combined the principles of non-maleficence and beneficence to become 'respect for wellbeing'.

The ethical matrix is an analytical tool and in itself is value neutral. It has the three principles (wellbeing, autonomy and justice) on the horizontal axis. On the vertical axis one lists the interest groups—that is, the people, organisations, communities, and so on—who stand to be affected by the decisions being made. The task then is to identify and document the ethical impacts of the matter under consideration in each cell of the matrix. While this task can be undertaken through desk-based research, it is also a dialogue tool when undertaken through group discussion. As Mepham (2000:168) clarifies in relation to introducing a new technology, 'actors (e.g., members of a regulatory committee) are asked to imagine themselves to be members of each specified interest group in turn, and to assess the ethical impacts of the introduction of the proposed technology'.

Once the cells of the matrix have been filled in, its users need to weigh the relative importance of the issues identified. Different people might give different weights to a given potential ethical impact on a particular interest group. Through discussion, the users of the matrix reach agreement about how the options under consideration, if implemented, will affect the different interest groups with respect to their wellbeing, autonomy and entitlement to justice. They seek to reach consensus.

As noted above, ethical matrices can be developed as a desktop exercise (for example, Food Ethics Council 2001) or as a participatory exercise with interest groups (the steps involved are detailed in Mepham et al. 2006). A mixed approach has also been used, in which subject-matter experts and ethicists develop the ethical matrix (that is, they fill in the boxes in the matrix) and then workshop it with stakeholders to produce '[a] practical consensus on the content of the matrix…followed by a weighing of the most important values' (Kaiser and Forsberg 2001:193).

The ethical matrix is a tool for identifying and analysing the potential ethical impacts of decisions. It is particularly applicable when decision makers bring to the table and/or have awareness of diverse values potentially influencing their decisions, and feel that these need to be integrated to produce a decision that is value informed and that has been attained in an explicit manner: 'While it might guide individual ethical judgements, the principal aim of the Matrix is to facilitate rational public policy decision-making by articulating the ethical dimensions of any issue in a manner that is transparent and broadly comprehensible' (Mepham 2000:169).

It deals head-on with the reality that, in decision making on important issues facing communities, 'some values seem to weigh heavier than others' (Kaiser and Forsberg 2001:193). Integrating stakeholders' values through developing an ethical matrix can reduce the potential for decision makers to ignore the values of the weakest actors.

Example of its use in research integration

Technological innovation: identifying the stakeholders of bio-remediation projects, and ethical issues involved in bio-remediation

What was the context for the integration, what was the integration aiming to achieve and who was intended to benefit?

Bio-remediation is a technological innovation in which micro-organisms and/or plants are used to locate, degrade or remove pollutants from the environment. Some people see it as a sustainable approach to dealing with environmental pollutants at a time when the use of landfill for the disposal of pollutants is becoming increasingly less sustainable. The UK Biotechnology and Biological Sciences Research Council commissioned a study into the feasibility of using these technologies in the United Kingdom using dialogue methods to focus on the potential social and ethical issues involved (Mepham et al. 2006).

What was being integrated?

The potential impacts on four 'stakeholder' groups—users of bio-remediation methods, affected citizens, technology providers and the environment—were examined. This was done through five focus groups: a non-governmental organisation (NGO), a national women's organisation, a technology/regulator group and two general public groups.

Who did the integration and how was it undertaken?

The ethical matrix tool was used in five focus groups. Each group used the ethical matrix to identify the potential impacts of bio-remediation for four stakeholder groups, taking into account the principles of wellbeing, autonomy and justice. The approach taken was: 'Participants considered whether the application of the technology might infringe [on] or respect the principles as applied to each of the interest groups. Participants were also asked to examine the types of formal and informal policies that might enhance respect for the ethical principles for the chosen interest groups' (Mepham et al. 2006:37).

The following ethical matrix resulted:

Respect for:	Wellbeing	Autonomy	Justice
Users	Efficacy, safety and remuneration	Freedom to adopt or not adopt	Fair treatment in trade and law
Affected citizens	Safety and quality of life	Democratic decision making	Individual and regional justice
Technology providers	Commercial viability and working conditions	Ability to innovate	Equitable trading (market) system
Environment	Protection of the environment	Biodiversity of biotic populations	Sustainability of the environment

Source: Mepham et al. 2006: 37.

Examples of the issues raised in group discussions were:

- with respect to the cell covering the principle of wellbeing and the interest group of affected citizens (that is, safety and quality of life): 'When building houses on bioremediated sites, concerns were raised that not all contamination may be "removed". Concerns [were] expressed regarding impacts on vulnerable groups (e.g. children) and possible risks from growing fruit and vegetables'
- with respect to the autonomy/environment cell (that is, biodiversity of biotic populations): 'Concerns [were] expressed regarding potential impacts on wildlife from phytoremediation (e.g. poisoning, bio-accumulation)' (Mepham et al. 2006:38–9).

What was the outcome of the integration?

Participants' deliberations about the values important to each of the four identified interest groups, set out in the cells of the matrix, resulted in a conservative orientation to the use of bio-remediation They generally took a precautionary approach rather than enthusiastically embracing the new technology. In other words, the integration of participants' values that occurred through the discussions, producing the agreed-on lists of ethical concerns illustrated above, led to the conclusion that the value issues or challenges involved in the innovation were formidable. Participants were generally satisfied with this tool, with more than 85 per cent stating that the ethical matrix contributed positively to the discussions (the balance expressed neutral views), helping them to attain group consensus on the ethical issues involved and their relative importance.

Commentary

The ethical matrix is a tool for integrating the values held by different stakeholders in an initiative, or for anticipating what their values might be and how they will be differentially impacted on by the options available for implementing the initiative. A particular strength is in raising the salience of values and value conflicts, and the importance of dealing with these in research integration. The method is grounded in people's own ways of seeing values, rather than using an imposed framework, and is conceptually straightforward.

The example provided goes beyond research integration to involve, in the values exploration and integration, a number of other entities—namely, an NGO group, a national women's organisation, a technology/regulator group and two general public groups. A narrower focus on research integration would have occurred if the participants were researchers, from various disciplines, using the method to reveal and explore the values underlying their individual practice and their disciplinary orientations. Similarly, the ethical matrix could have been completed through dialogue between researchers and decision makers, achieving similar goals of transparency about values and their impacts on evidence-based decision making.

While the surfacing of values (filling in the cells of the matrix) is a strength of the method, a weakness is that it does not provide any clear guidelines for how to move towards a consensus on the values so identified. Methods for discussing and reaching agreement on the relative importance of the opposing values brought to the surface, and on the relative impacts on the different stakeholders of the initiative being examined, are not included in the process. Instead, users rely on skilled facilitators to guide this process and assist the participants to find consensus through group discussion.

There are probably occasions in which people could use the ethical matrix effectively to identify important stakeholders and to elicit the value issues important to them, but are unable to reach consensus on the relative weight of the conflicting values and the implications for action flowing from this. When much diversity exists between participants—for example, in an interdisciplinary research team or a situation in which researchers, policy people, practitioners and affected communities are involved—significant challenges may exist in finding consensus on values.

It is possible that using another dialogue method to integrate the judgments of participants (or of another group of stakeholders) about the relative importance of the values brought to the surface and of the likely impacts of the intervention, could follow the ethical matrix exercise. The Delphi technique or the nominal group technique would be suitable for this task.

Origins and genealogy

This tool was introduced by Professor Ben Mepham, Special Professor in Applied Bioethics at the Centre for Applied Bioethics at the University of Nottingham, in 1994. Since then it has been applied and modified in various settings, focusing on a range of ethical issues, many of which deal with food security and natural resource management. The UK Food Council (<http://www.foodethicscouncil.org/>), of which Mepham is a member, has used the ethical matrix tool extensively and promotes its further development and application.

Further reading on the ethical matrix

Food Ethics Council n.d., *Ethical Matrix*, Food Ethics Council, <http://www.foodethicscouncil.org/ourwork/tools/ethicalmatrix/introduction>

Mepham, B., Kaiser, M., Thorstensen, E., Tomkins, S. and Millar, K. 2006, *Ethical Matrix Manual*, LEI, The Hague, <http://www.ethicaltools.info/content/ET2 Manual EM (Binnenwerk 45p).pdf>

Schroeder, D. and Palmer, C. 2003, 'Technology assessment and the "ethical matrix"', *Poiesis & Praxis: International Journal of Technology Assessment and Ethics of Science*, vol. 1, no. 4, pp. 295–307.

Endnotes

[1] This case example is not structured around the standard six questions (Appendix 1) owing to the characteristics of the dialogue method and the limitations of the source material on which it is based.

Chapter 5. Differentiating between the dialogue methods

We have presented 14 dialogue methods that can be used to structure research integration. These are group processes to jointly create meaning and shared understanding about real-world problems by bringing together knowledge from relevant discipline experts and stakeholders. Ten are methods for creating broad understanding about a problem; they integrate the participants' judgments. Four are specific methods that can drill down into a particular aspect of a problem that might be contentious or of particular significance. The latter methods examine participants' visions, world views, interests and values.

As far as we are aware, this is the first time that dialogue methods have been explored specifically for their value in research integration. Our primary aim is to broaden the range of methods available to researchers who already have some experience in research integration. Consequently, our focus is on describing the methods and providing real-world examples of how they have been used.

Future research could valuably start to differentiate between methods, so that research integrators can easily distinguish which method is best suited for a particular integration purpose. We start this process with an exercise on a real-world problem—changes in illegal amphetamine use—using hypothetical dialogue questions. For each of the 14 methods in the book, we describe a key question related to the amphetamines problem that the method is particularly well equipped to handle. We also describe the discipline experts and stakeholders who would typically be drawn on to address such a question in a research integration context. This by itself is already useful in starting to demonstrate differences between the dialogue methods. We extend this by cross-tabulating the 10 dialogue methods for broad understanding, with the 10 key questions we have developed for them. We then look at each method against each question to determine which of the other methods are likely to be useful for addressing each question.

The problem we have chosen to focus on is the use of illegal amphetamines by young people. While the application of this example in the 14 dialogue methods is hypothetical, the problem we describe is real. A brief synopsis is as follows. In recent years, there have been challenging changes in the patterns of use of amphetamines in Australia, with a move away from powdered amphetamine ('speed') to a potent crystalline form of methamphetamine ('ice'). In addition, there has been a transition from oral ingestion to injecting. Key issues for relevant stakeholders are as follows:

- For users who engage in high-frequency and high-speed injecting, there are likely to be problematic health, social and financial consequences, including acute psychotic-like episodes accompanied by violence, the development of dependence, difficult withdrawal symptoms with agitation and depression, and stress on relationships.
- For treatment providers, these clients are often difficult to manage, especially when they are violent, agitated, hypersensitive and unable to concentrate. Treatment options are limited to cognitive-based therapies, with no pharmacotherapies (that is, drug treatments) available.
- For police, ambulance officers and hospital emergency workers, the violent behaviour of users can be a major problem, especially as force and administration of morphine are the most commonly used ways to calm them down.
- For drug user organisations (also known as peer-based organisations), there is an important role in developing and distributing advice on how to reduce harm, including information about safer injection practices, concomitant drug use, safe sex and so on. Peer-based outreach workers, who seek out amphetamine users, can be an important part of such strategies.
- For police, the drug sources include local manufacture and importation. Clandestine local laboratories pose risks of explosion and fire. The drugs are easy to conceal for importation. There is little evidence that police interdiction ('busts') involving significant amounts of these drugs have any impact on their availability.
- For pharmaceutical companies and pharmacists, the constituents of legal drugs (pseudoephedrine) are the precursor for illicit amphetamine manufacture. Depending on the scale of the illegal operation, pharmaceutical companies can be targeted for precursors or pharmacists can be approached to obtain legal drugs from which precursors are then extracted.
- For society in general, there is a false perception of widespread use, which can encourage normalisation of this problematic behaviour. Reporting is often seriously exaggerated and concern about the adverse consequences can be out of proportion. For example, amphetamine users are less likely to die than heroin users.

Discipline-based researchers have substantial contributions to make to understanding these problems, for example:

- assessment of the prevalence of amphetamine use and of the various harms, as well as the characteristics of those most likely to be affected (epidemiologists)
- identification of medical problems caused by or associated with amphetamine use (clinical researchers)
- evaluation of treatment options (clinical researchers)
- examination of drug markets and the impacts of various law enforcement strategies (criminologists)
- investigation of the different behaviours associated with amphetamine use and how violent behaviour, for example, might be ameliorated (psychologists)
- detailed observation of the lives of amphetamine users (ethnographers)
- understanding the causes of amphetamine use (psychologists, sociologists and/or epidemiologists)
- calculating the treatment and law enforcement costs of amphetamine use (economists)
- investigating the social costs of amphetamine use (sociologists).

We now describe a characteristic research question about this problem that each dialogue method is well suited to address (Table 5.1). We also suggest a typical array of discipline and stakeholder experts whose knowledge about the problem will contribute to each form of dialogue and who can be expected to be included among the participants.

Table 5.1 Characteristic research questions for each of the dialogue methods, plus typical discipline and stakeholder participants in a research integration process

Dialogue method	Characteristic research question	Typical participants (disciplines/stakeholders)
Broad methods		
Citizens' jury	Is amphetamine use a priority for community action?	(Discipline experts provide information)
		Cross-section of the public
Consensus conference	How can the community best respond to amphetamine use?	(Discipline experts provide information)
		Cross-section of the public
Consensus development panel	What are the best-practice guidelines for treatment of amphetamine users?	Clinical researchers
		(Unlikely to have stakeholder representatives, although the clinical researchers will generally also be treatment providers)
Delphi technique	What is the nature and extent of harm arising from amphetamine use?	Clinical researchers, epidemiologists, psychologists, sociologists
		(Might not have stakeholder representatives, although families, police, treatment providers and users could be included)
Future search conference	What is the future of young people in a society with high availability of stimulant drugs?	Clinical researchers, epidemiologists, psychologists, sociologists
		Churches, families, media, police, schools, treatment providers, users
Most significant change technique	What are the outcomes of peer education among 'ice' users?	Ethnographers
		Peer educators, users
Nominal group technique	How can police, ambulance officers and emergency workers better respond to acute psychosis and violent behaviour among amphetamine users?	Psychologists
		Police, ambulance officers, emergency medicine specialists
Open space technology	How can the harms from high-frequency, high-risk injecting best be reduced?	Clinical researchers, ethnographers, psychologists, sociologists
		Peer educators, police, treatment providers, users
Scenario planning	What is the best balance between direct (eg., drug seizures) and indirect (eg., precursor control) law enforcement methods?	Criminologists
		Criminal intelligence analysts, police, pharmaceutical company representatives
Soft systems methodology	What are the key considerations for a national government action plan on amphetamines?	Clinical researchers, epidemiologists, psychologists, sociologists
		Churches, families, media, police, policy-makers, schools, treatment providers, users

Table 5.1 (continued)

Dialogue method	Characteristic research question	Typical participants (disciplines/stakeholders)
Specific methods		
Appreciative inquiry	How can a busy hospital emergency unit best deal with amphetamine users?	Clinical researchers, psychologists, sociologists
	(Different visions are likely to be important)	Emergency medicine specialists, users
Strategic assumption surfacing and testing	How should a treatment service respond to violent users?	Clinical researchers, psychologists, sociologists
	(Different world views are likely to be important)	Treatment providers, users
Principled negotiation	How can pharmacies, police and government best cooperate on precursor control?	Criminologists
		Pharmacists, police, policy-makers
	(Different interests are likely to be important)	
Ethical matrix	Should schools suspend amphetamine users?	Criminologists, education researchers, psychologists, sociologists
	(Different values are likely to be important)	Parents, police, school principals, students, teachers, users

The 10 research questions, which are typical of those addressed by the individual methods to gain a broad understanding, are:

1. Is amphetamine use a priority for community action?
2. How can the community best respond to amphetamine use?
3. What are the best-practice guidelines for treatment of amphetamine users?
4. What is the nature and extent of harm arising from amphetamine use?
5. What is the future of young people in a society with high availability of stimulant drugs?
6. What are the outcomes of peer education among 'ice' users?
7. How can police, ambulance officers and emergency workers better respond to acute psychosis and violent behaviour among amphetamine users?
8. How can the harms from high-frequency, high-risk injecting best be reduced?
9. What is the best balance between direct (for example, drug seizures) and indirect (for example, precursor control) law enforcement methods?
10. What are the key considerations for a national government action plan on amphetamines?

In Table 5.2, we cross-tabulate the 10 methods for broad understanding and the 10 questions.

Table 5.2 Which methods for broad understanding are well suited to answer each of the characteristic research questions?

Method	Question									
	1	2	3	4	5	6	7	8	9	10
Citizens' jury	√	√								√
Consensus conference	√	√								√
Consensus development panel		√	√	√			√	√	√	
Delphi technique			√	√			√	√	√	√
Future search conference		√			√					√
Most significant change technique						√		√		
Nominal group technique			√	√			√	√	√	√
Open space technology	√	√			√			√		
Scenario planning		√			√				√	
Soft systems methodology		√			√		√	√		√

This exercise demonstrates that most dialogue methods are suitable for more than one type of question, but also that different dialogue methods are particularly applicable for answering different types of questions, and for doing so in different circumstances. The domains in which the individual methods are particularly applicable become clearer when we look at the questions one at a time, noting which methods fit best, and the reasons for this.

1. Is amphetamine use a priority for community action?

As well as citizens' juries, the consensus conference and open space technology were classified as being appropriate dialogue methods for dealing with this question. This reflects its emphasis on the community as a key stakeholder, hence the need for a method that taps community—rather than expert—assessments. It is noted, however, that the three methods include expert inputs, frequently from researchers, to assist the citizens to make informed judgments.

In contrast, the scenario planning method is inappropriate as the task does not include developing scenarios. Soft systems methodology is not appropriate as it is based on a shared understanding, from the outset, that a problem exists (rather than exploring the seriousness of the problem, as here) and has a distinctly action-oriented focus. The breadth of the question, and the need to tap informed community, rather than expert, judgments, means that the Delphi technique is less appropriate than the nominated methods.

2. How can the community best respond to amphetamine use?

In addition to the consensus conference, most of the dialogue methods listed have been assessed as suitable for responding to this question—the exceptions being the Delphi technique, most significant change technique and the nominal group technique, which are assessed as being unsuitable. This reflects the breadth of the question and the fact that a range of stakeholders—for example, community members, experts, decision makers, and so on—are able to contribute to finding answers to it.

The in-depth exploration by stakeholders that results in action plans, as occurs in soft systems methodology, is apposite here. As with question one, the breadth of the question means that methods such as the citizens' jury, consensus conference and open space technology are highly suitable as all are useful in opening up the issues, exploring a variety of possible responses and reaching judgments on the most appropriate responses. This breadth means that the Delphi technique and nominal group technique are less appropriate. The most significant change technique is unsuitable owing to its focus on understanding outcomes in the context of evaluation—a consideration not relevant here.

3. What are the best-practice guidelines for treatment of amphetamine users?

This was the typical question for the consensus development panel, but the Delphi technique and the nominal group technique would also be suitable. This is because the question needs to be answered by experts and, since difference of opinion is likely to exist among experts on the topic, a highly structured method is needed to tap their judgments and synthesise them with those of their peers. The narrowness of the question is also an important consideration. A fair degree of control of the process is needed to produce results, in contrast with other, more open, free-flowing dialogue techniques.

The methods that are designed to elicit the judgments of citizens, rather than experts, are inappropriate here owing to the subject matter. The most significant change technique is irrelevant as it is not a program or policy evaluation task, and scenario planning is also unsuitable as eliciting and weighing current knowledge is the focus, not developing scenarios for the future.

4. What is the nature and extent of harm arising from amphetamine use?

As with question three, we have assessed the consensus development panel, the Delphi technique and the nominal group technique as being the dialogue methods best suited to answering this question. This reflects the need for expert assessments and the narrowness of the question.

5. What is the future of young people in a society with high availability of stimulant drugs?

This question was designed for the future search conference method, but three other methods could also be helpful in finding answers: open space technology, scenario planning and soft systems methodology. Scenario planning techniques could be used to develop a range of different scenarios given different assumptions about such things as the availability of amphetamines, patterns of use, population groups with high prevalence of use, societal responses, and so on. This detailed scenario development could build on a more inclusive method, such as open space technology, taking its products as inputs to scenario planning.

The citizens' jury and consensus conference methods are not well suited in this case owing to their lack of focus on the future. The question is not in the realm of evaluation so excludes the most significant change technique. It is too broad for the consensus development panel, Delphi technique and nominal group technique and calls for inputs from a range of stakeholders, not just experts.

6. What are the outcomes of peer education among 'ice' users?

Uniquely among the 10 questions, only the dialogue method for which this question was developed—the most significant change technique—has been identified as particularly useful in addressing it. This is because the core function of the most significant change technique is to contribute to evaluation, especially program evaluation. It is a narrative technique to elicit the stories that best illustrate the most important outcomes of a program. Clearly, these outcomes could be positive, negative or a combination of the two, so the process of eliciting them needs to be sensitive to the possibility of bias towards surfacing the positive outcomes and concealing the negatives.

The emphasis on summative and outcome evaluation in the question distinguishes it from the others and makes dialogue methods other that the most significant change technique either far less suitable than this method or completely unsuitable.

7. How can police, ambulance officers and emergency workers better respond to acute psychosis and violent behaviour among amphetamine users?

Four methods seem useful for addressing this question: the consensus development panel, the Delphi technique, the nominal group technique and soft systems methodology. This reflects the fact that, to answer the question, people with substantial knowledge and experience of the topic need to be involved, meaning that methods based on tapping citizens' judgments are excluded. A high degree of structure in implementing the method, with the locus of control found in the organisers and facilitators rather than participants, is important, and is found in these four methods. It is output oriented, rather than process oriented. The question is complex, addressing areas of uncertainty, meaning that soft systems methodologies will be useful.

The most significant change technique could also be applicable—although probably not as directly as the other methods listed—as an evaluative element exists. Narratives demonstrating sound outcomes when police, ambulance officers and other emergency workers use certain approaches to dealing with amphetamine users exhibiting violent behaviour could be generated and assessed using the most significant change technique.

The narrowness of the question excludes the broad exploration of issues that characterises the future search conference and open space technology.

8. How can the harms from high-frequency, high-risk injecting best be reduced?

Although we developed this question as typical for open space technology, the consensus development panel, Delphi technique, nominal group technique and soft systems methodology found helpful in addressing the previous question are also applicable here. As in the previous question, what they will bring are highly structured methods able to tap the knowledge of experts in addressing an area characterised by uncertainty. The output would be agreed strategies and action plans to implement them.

We originally envisaged the question for an open space technology event in which most of the participants were people who used illegal amphetamines. We expect that they could produce new insights and action plans, grounded in their lived experiences, which they and others could implement.

The most significant change technique is also potentially useful, though perhaps not as useful as the other five. The narratives produced by the most significant change technique, demonstrating sound outcomes from particular strategies aiming to reduce the harms associated with high-risk injecting, could inform the development of strategies and action plans.

Methods giving high salience to tapping public opinion and judgments, especially the citizens' jury, consensus conference and future search conference, are inappropriate in this case.

9. What is the best balance between direct (for example, drug seizures) and indirect (for example, precursor control) law enforcement methods?

This is another question that could be dealt with effectively by a number of dialogue methods. We developed the question for scenario planning as different scenarios could be developed for the two broad strategies, facilitating comparison of their utility in contributing to planning. The three highly structured methods for tapping expertise and finding agreement in the face of uncertainty (the consensus development panel, the Delphi technique and the nominal group technique) would work well here as experts have knowledge on the topic to bring to the dialogue.

The question is too narrow for a future search conference or the open space technology, and calls for a more structured approach than used in these methods. Since evaluation is not its focus, the most significant change technique also has limited application here.

10. What are the key considerations for a national government action plan on amphetamines?

This is a typical question for soft systems methodology, as a need exists to understand the whole picture, to set boundaries for the action plan and to understand the implications of leaving some aspects out of the scope of the action plan. The process needs to be inclusive, structured and focused on the product: an action plan that all participants/key stakeholders are willing to sign up to.

The question could also be dealt with using other dialogue methods, with their somewhat different focuses. As with question nine, the methods for tapping expert opinion and helping experts reach consensus on the key considerations are useful, especially the Delphi technique and the nominal group technique. Since there is no single, clear answer to be found in expert knowledge, individual judgment would be important. For this reason, the methods that minimise the impacts of power differentials between the expert participants (the Delphi technique and the nominal group technique) would be better than the face-to-face, round-table discussion approach used in the consensus development panel.

Again, because there is no single best answer available to the question, and the community at large is a stakeholder, informed citizens could make positive contributions to answering it. Hence, citizens' juries and the consensus conference, being designed and implemented to ensure that the involved citizens are well informed of the options, along with their strengths and weaknesses, could be particularly helpful. The future search conference method would also work, provided the question was expressed differently, perhaps as 'The future of government action on illegal amphetamines: what are the key considerations?'.

The low salience of evaluation here excludes the most significant change technique; the lack of certainty in the evidence base and the need for dealing with power differentials between various experts excludes the consensus development panel method; and the absence of a need to develop scenarios excludes the scenario planning method. Open space technology could be useful in generating ideas, but its unstructured approach means that its products would probably be diffuse, reflecting the areas of interest of the most influential participants rather than a well-balanced, comprehensive exposition of the key considerations.

Comments

In this exercise, we have described a situation in which there are serious problems consequent on the availability and use of amphetamine-type substances, and on societal responses to these. We presented 10 research questions that could arise in such a context, and discussed how the various dialogue methods could be used to address each of them, highlighting those that would be most apposite, those less so and those unsuitable for that particular purpose. We have shown that it is generally not possible to make hard and fast pronouncements, as many of the methods are flexible and adaptable. In addition, the questions can also often be addressed in different ways, emphasising different aspects of the question. Nevertheless, it is also clear that all the methods are not equally suitable for all of the questions.

We have started to identify criteria that differentiate between the methods, including:

- the narrowness or breadth of the research question
- the level of complexity in the research question
- the balance between empirical facts and subjective judgments
- the types of participants engaging in dialogue—for example, citizens versus subject-matter experts
- the degree to which the methods deal with power differentials among the participants
- the desirability or otherwise of face-to-face engagement
- whether a specific purpose is to be filled—for example, evaluating a program or developing scenarios.

Further differentiating between the dialogue methods, to provide guidance to research integrators in their use, is an important task for future research. Particular benefit will be derived from researchers documenting their experiences in using dialogue methods in research integration and evaluating the outcomes, along with the factors most salient in producing successful outcomes. Some of these factors will be intrinsic to the method (for example, face-to-face versus anonymous), some will be dependent on the research question addressed (for example, the narrowness or breadth of the question), some will reflect contextual factors (for example, the auspices under which the dialogue was conducted) and yet others will rely on the skills of the personnel using the dialogue method. The development of such a body of knowledge is likely to allow research integration specialists to work towards creating a decision tree to guide people in selecting the most appropriate dialogue method to attain their goals, given their situation and constraints.

Chapter 6. Conclusions

Learning from failure

In our search for case studies, we found only one description of a failure. This was in a situation in which researchers used the nominal group technique to try to change land managers' attitudes and values (Padgett and Imani 1999). Specifically, they aimed to move land managers to a position from which they would be more accepting of US Government policies on environmental justice. The nominal group process, however, had the opposite effect, moving them to an even more conservative position than the one they held before the group experience. As the authors explain, the most likely reason is that the participants did not want to be there at all and therefore rejected the dialogue process.

This highlights an issue that we have not raised so far—namely, that dialogue methods rely on the willing participation of all involved. This is an unstated assumption behind all of the methods we describe in this book. This also points to a weakness of dialogue as a research integration process: if some key disciplinary or stakeholder groups, or sections of a particular group, do not want to be involved, the dialogue process becomes skewed and its usefulness can be limited.

The publication of such negative findings, critical analysis of individual methods and comparisons between methods are all essential for determining the full potential and limitations of the use of dialogue methods in research integration.

Other research areas

We round off this conclusion by highlighting four additional suggestions for further research:

1. development of new dialogue methods or the extension of existing ones to address other aspects of research integration
2. continued cross-fertilisation between areas such as natural resource management, public health, security and technological innovation in methods development
3. exploration and documentation of flexible, and even improvised, combinations of methods
4. improvement of dialogue methods through critical analysis and evaluation.

We are not aware of other investigations that have tried to link research integration elements and dialogue methods and have been pleased to discover through our work that this is an area that has promise. We were struck by the potential of the dialogue methods we examined to focus on and strengthen particular aspects of research integration. As we discussed in chapter 5, more work is needed to tease out which methods are most suitable for particular research integration tasks. In addition, there are areas—such as the integration of visions, world views or interests—where we have been able to identify only one method and where the development of additional techniques is likely to be valuable. Furthermore, our analysis of the elements of knowledge to be integrated was simple and pragmatic and remains open for more sophisticated development.

We note that, with the exception of the Delphi technique, none of the other methods we describe could be illustrated by examples in every one of the four areas of application that we chose to investigate: the environment, public health, security and technological innovation. While this could be an artefact of our search strategy, we suggest it is more likely that formal dialogue methods are not yet exploited to their full potential in research integration. Cross-fertilisation between these four (and other) areas is likely to have at least two substantial benefits. First, it will draw the attention of researchers in a particular area, such as security, to the potential of methods new to that area. This not only increases the methods repertoire of security researchers, it could stop them from reinventing the wheel if they decide a new method is needed. Second, documentation of experiences in different areas could provide insights into useful modifications in applications of particular methods, as well as specific dos and don'ts that are key to the method's success.

One of the other benefits of cross-fertilisation is that it can alert researchers to the importance of flexibility in how methods can be used singly and in combination. We pointed out the importance of flexibility in our introductory sections and re-emphasise it here. The hallmark of excellent experienced research integrators specialised in dialogue methods[1] will be their ability to mix and match methods as the needs of a particular research problem require. This was drawn to our attention by two of the people we asked to review an earlier draft, Gerald Midgley and Wendy Gregory, whose systemic intervention practice (see Midgley 2000) illustrates these principles. This could be taken even further to an appreciation of the value of improvisation once a dialogue process is in full swing. While preparation for dialogue events is essential, surprises can occur once the group convenes. Flexibility and improvisation are important not only in choosing and adapting particular methods, but in the areas of making groups 'work', which we do not deal with specifically in this book. One of the key lessons provided by the teaching of improvisation to jazz students is the importance of making explicit knowledge tacit (Bammer and Smithson 2008; Mackey 2008) so that when the time comes to 'take a solo', the performer can organically draw on that store of knowledge. As Mackey (2008:107) has described, the ability to improvise in jazz is built by internalising explicit and performance knowledge about 'accentuation, articulation, tempo, ornamentation, rubato, melody and rhythm'. We suggest that flexibility and improvisation are also areas for further development in research integration using dialogue.

Recognition that dialogue is an art should, however, not diminish attention to the twin aspect of dialogue as a science. In the description of the failure that we present above, it is clear that dialogue is not a case of 'anything goes'. There will be parameters that determine success or failure and teasing those out is critical to the future of dialogue as an effective research integration method. Two important areas to explore are bias through participant selection and maximising the benefits of conflict.

With regard to the first of these, there is need for greater consideration of the potential for bias through the selection of participants. While some dialogue methods, such as the citizens' jury, try to ensure that their participants are as representative as possible of the population of interest, the potential for bias seems not to be considered in the application of other methods.

The second area involves appreciating the importance of conflict and different strategies for maximising the benefits of conflict (and minimising the costs). It is likely to be an implicit assumption that dialogue should lead to consensus, but in fact that is not necessarily achievable, let alone what should be aimed for. A few of our examples (such as the future search conference on reducing the human and economic costs of RSI and the use of strategic assumption surfacing and testing in a US Cooperative Development Agency) demonstrate non-consensual outcomes, and there is substantial scope for further investigation of this area.

An invitation to contribute

Researching and writing this book has made us even more enthusiastic about the potential of dialogue methods as research integration tools to more effectively tackle real-world problems.

This book provides a compilation of available methods and cases of their application to problems in the environment, public health, security and/or technological innovation. We strongly encourage those involved in using dialogue for research integration to publish their findings—not only in terms of the outcomes, but in terms of the processes used. We hope that our six-question framework (see Appendix 1 and the cases above) will provide a useful way of structuring such publications. Based on an analysis that one of us has been involved in (Kueffer et al. 2007), we appreciate that outlets for such publications are limited. Therefore, for those who have difficulties in finding a suitable journal or other place to publish, especially for lessons learnt from processes that have not gone well, we have established a place on the Integration and Implementation Sciences web site (<http://i2s.anu.edu.au/>) that can be a repository for such cases and that provides an opportunity for discussion of dialogue as a research integration method, as well as a site for recording feedback on this book.

Endnotes

[1] While we would expect all those trained in Integration and Implementation Sciences to be familiar with dialogue methods for research integration, we would not expect everyone to be accomplished in running and facilitating such dialogues. This could be an area for specialisation within the crosscutting discipline.

Appendix 1. Dialogue methods in the context of Integration and Implementation Sciences

Gabriele Bammer

The aim of this book is to provide a specific methodological 'tool kit' for researchers who focus on real-world problems and who seek to bring together disciplinary and stakeholder insights into a particular problem. This tool kit focuses on dialogue methods for bringing together multiple perspectives to address real-world problems.

As we foreshadowed in the introduction, we see this as only one of the methodological skills that researchers oriented towards real-world problems require. We argue for a particular set of conceptual and methods skills, which we call Integration and Implementation Sciences (I2S). Here, we describe the core elements of this new crosscutting discipline. We have provided details about the rationale behind I2S elsewhere (Bammer 2005, 2008b), so this is presented only briefly here. We then outline the four cornerstones of I2S. We focus on one of those cornerstones—integrating disciplinary and practice (stakeholder) knowledge—where the dialogue methods for research integration are located, to put this tool kit into a broader context.

Rationale for developing I2S

There is growing appreciation of the importance of interdisciplinary and trans-disciplinary research that focuses on real-world problems (referred to from now on as cross-disciplinary research), alongside research that advances understanding through a single discipline. There are three challenges in conducting such research that are not yet resolved:

1. there are no well-established institutional structures within which to undertake real-world research
2. there is no accepted way to deal with weaknesses in current approaches to cross-disciplinary problem-based research
3. recurrent issues in tackling real-world problems that are not within the domain of any discipline or practice area.

I2S aims to provide a solution to all three challenges. In terms of institutional structures, I2S shifts the focus away from the content of real-world problems to the methods for addressing them. In other words, rather than trying to find agreed ways to institutionalise approaches to specific problems, such as multitudes of centres covering bio-security, climate change, obesity, tobacco control, and so on, approaches to cross-disciplinary problem-based research are institutionalised through a discipline of Integration and Implementation Sciences, which will be accommodated as a standard academic department. I2S researchers then provide the linchpin for investigations into real-world problems, providing a concrete hub around which diverse discipline and practice perspectives can be drawn on in a flexible manner. The discipline and practice experts involved in investigating the real-world problem can change as the requirements of the investigation change.

In terms of how I2S will be organised, some useful analogies can be drawn with statistics. Like statistics, I2S is a crosscutting discipline that works on three levels: i) a core of people, the disciplinary specialists, who focus on the development of integration and implementation theory and methods; ii) a substantially larger group of people, the equivalent of applied statisticians, who focus on integration and implementation in relation to specific problem areas—for example, in bio-security, the environment, population health, and so on. They not only apply what is known to the problem area, they use their work on the problem to develop new integration and implementation theory and methods; and iii) just as most researchers have at least a basic appreciation of statistics, its uses and where to find advanced expertise when they need it, most researchers will also have a similar appreciation of Integration and Implementation Sciences.

A disciplinary hub focusing on integration and implementation theory and methods also addresses two fundamental weaknesses that cross-disciplinary problem-based research suffers from as it is currently conducted. The first is that, unlike discipline-based research, there is no core methodological underpinning to cross-disciplinary problem-based research. As a consequence, the quality of any particular piece of such problem-based research is not guaranteed and is hard to assess. Furthermore, the outcomes are likely to be much more hit-or-miss than in discipline-based research. The second weakness is that while many cross-disciplinary problem-based teams develop new concepts and techniques, there is no recognised systematic way of communicating such insights between problem-based teams. In this book, we see this with dialogue-based methods for research integration, with the communication gap between disparate problem areas, so that, for example, researchers working on a problem of environmental management are unlikely to be aware of relevant innovations in public health or security. This has slowed progress in the development of cross-disciplinary problem-based research and has led to considerable 'reinvention of the wheel'. I2S aims to overcome this problem by providing cross-disciplinary problem-based researchers with a foundation of agreed core concepts and methods for undertaking their work, as well as an institutional mechanism for building the discipline and for communicating new developments.

The third area involves the domains that I2S should cover. We suggest that these should be areas relevant to research integration and implementation, which are not the territory of any established research discipline or practice area. The foundation of I2S is therefore twofold. On the one hand, it involves compiling concepts and methods developed in disciplines or practice areas that individually address only part of the problem. In addition, it involves bringing together ideas and techniques developed in addressing a particular problem, for which there is currently no communication mechanism, as outlined above. This book of dialogue-based integration methods is an example of the latter. A recent book bringing together a range of disciplinary and practice perspectives on uncertainty (Bammer and Smithson 2008) is an example of the former. In other words, I2S covers integration and implementation concepts and methods that none of the established disciplines or practice areas has the mandate to pull together.

The four cornerstones of I2S

We suggest that there are four domains that I2S covers. Real-world problems require not only the integration of insights from diverse discipline and practice perspectives, as presented in this book, they need new thinking to determine ways forward, they require effective management of knowledge gaps and uncertainties, and they need effective uptake of research findings into policy and practice change.

We present this diagrammatically, as follows:

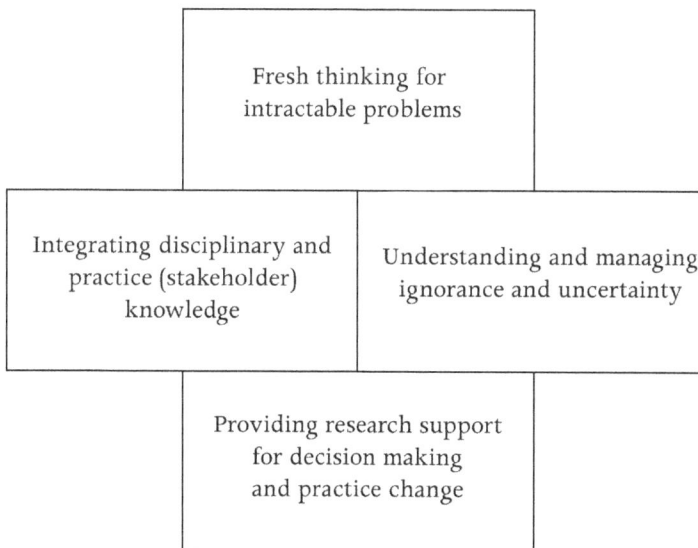

Many of the real-world problems that societies face are intractable, so that sparking innovative thinking about them is essential. For example, how do we balance the rights of individuals with the prevention of abuse of legal safeguards by criminals; how do we motivate young people to become engaged, productive citizens; how do we encourage independence in medical research but restrict the development of potentially dangerous viruses? I2S seeks to develop concepts and methods that can catalyse innovative ways forward in thinking about such problems, leading to more effective policy and practice approaches.

Integrating disciplinary and practice (stakeholder) knowledge has three elements: useful concepts, a range of effective methods and a standardised way of describing such integration. The dialogue-based methods we present in this book represent one class of methods. We describe this domain in more detail below.

Real-world problems also have many different types and aspects of ignorance and uncertainty embedded within them and there is currently no systematic way of recognising and dealing with all these attributes. Managing unknowns is just as important as making maximum use of what is known. This involves concentrated attention on the nature of ignorance and uncertainty, including the irreducibility of some uncertainties. It also involves understanding how people think about and respond to uncertainty—for example, through exploration of the metaphors they use, their motivations and even their moral orientations. Further, it involves examining different ways of coping and managing under uncertainty, especially in relation to meeting the adaptive challenges posed by uncertainty. The possibilities range from outright denial or banishment to acceptance and even exploitation of uncertainty. Each kind of response can be shown to have strengths and weaknesses that indicate when it is likely to be adaptive. While different disciplines and practice areas have established ways of dealing with ignorance and uncertainty—for example, statisticians focus on probability-based approaches, intelligence analysts focus on distortion, historians take taboo into account and psychologists think about norms—no discipline or practice area has the role of bringing all of these different approaches together (Bammer and Smithson 2008).

In terms of providing research support for decision making and practice change, we suggest treating decision making and practice change separately. For the former, in the past decade or so, there has been growing interest in the lack of impact much research has on policymaking and how this can be remedied. This is a subset of a larger problem—namely, how to increase consideration of research knowledge in decision making more generally, not only by policy-makers, but by business leaders, community activists, non-government organisations and professional groups. We suggest that this has five elements: a) understanding decision-making processes—for example, government policymaking or business commercial decision making; b) appreciating the attributes of influential research; c) delineating different types of researcher–decision maker engagement—ranging from one-way communication to the co-production of knowledge—and their strengths and weaknesses; d) understanding how institutions can influence what research is taken up by decision makers; and e) developing more effective ways to evaluate research support for decision making.

Furthermore, improving decision making might not necessarily lead to change on the ground. Understandings about how change occurs are widely dispersed in areas such as the diffusion of innovation, advertising, agricultural extension, health promotion, social entrepreneurship, community organising, organisational change and counselling. Again, no discipline or practice area has the mandate to bring all these perspectives together so that more can be learnt from the synergies and points of difference. Consequently, this is a key task for I2S.

Focusing on integrating disciplinary and practice (stakeholder) knowledge

As we outline above, integrating disciplinary and practice (stakeholder) knowledge has three elements: useful concepts, a range of effective methods and a standardised way of describing such integration. The dialogue-based methods we present in this book represent one class of methods.

The concept at the core of this domain is systems-based thinking. Systems thinking plays an important role in identifying interconnectedness. We need better approaches to understanding and managing connectedness to complement strong reductionist research methods. Reductionist research helps us understand single issues more deeply. It is key, for example, to identifying viruses or understanding group behaviour in panic situations. This needs to be balanced by systems thinking, which helps us deal with real-world problems such as responding to a terrorist attack or understanding the impact of a new epidemic.

Effective systems-based thinking plays out through ideas and, especially, methods to:

- improve scoping, problem framing and boundary setting, which define how a real-world problem will be approached and which perspectives will be included
- integrate effectively—for example, at the end of a multidisciplinary process or throughout a trans-disciplinary process
- identify and manage conflicts expediently between, for example, values, interests and epistemologies.

At present, each of these groups of methods is tackled on an ad hoc basis by cross-disciplinary problem-based researchers and, as indicated above, no discipline or practice area can fully inform their considerations, or act as a repository for the insights generated in various problem-based projects.

Scoping, problem framing and boundary setting are interlinked and are essential for more comprehensive approaches to problems and for making the inherent limitations of all cross-disciplinary problem-based research evident. In particular, it is impossible to research everything, let alone everything at once, so that the focus of a problem has to be restricted. Scoping, problem framing and boundary setting attend to different aspects of this.

As Bammer (2006c:4) has described elsewhere: 'Scoping is the preparatory stage of a project where we systematically think about what we can best do with the time, money, and people we have at our disposal in order to use those resources most effectively'. In terms of problem framing, the way we see problems and the language we use to describe them can play a powerful role in setting the basis for research integration. For example, research on terrorism could be defined or framed as 'an examination of individual factors involved in producing terrorist acts' or alternatively as 'an examination of cultural and environmental issues that lead to the commission of terrorist acts'. Both are about understanding why people engage in terrorist activities, but one approach frames it as a problem of individuals, whereas the other treats it as a societal problem, especially examining culture and the social environment. The way a problem is framed already implicitly sets some boundaries around the problem. The boundaries specify what will be included, excluded and marginalised (Midgley 2000). An important aspect of this for research integration is determining which disciplines and which non-academic or practice perspectives will be included in the project.

Second, integrating more effectively involves the development of a range of methods. This book describes a significant class of methods—namely, those based on dialogue. We suggest that there are four other classes of methods: model based, product based, vision based and common metric based (Bammer 2006d).

While modelling is not necessarily integrative, the process of building a model can be an effective way of bringing together different discipline and practice insights. For example, the Australian Cooperative Research Centre (CRC) for Greenhouse Accounting developed a modelling shell to encompass insights from soil science, plant biology and ecosystem dynamics. There is a wide array of modelling methods, ranging from conceptual mapping (Trochim and Trochim 2007) and formal system dynamics models (Maani and Cavana 2007) to agent-based models (Srbljinovic and Skunca 2003) and purpose-built models such as those used in the CRC for Greenhouse Accounting example. The utility of the final model can be a measure of how well the integration has been conducted.

Because modelling methods are used so widely, we class them in their own right, but they can also be seen as a special case of product-based methods. Like building a model, developing a product can be effective for synthesising discipline and practice knowledge. Probably the best-known example is the building of the atomic bomb. The atomic bomb project brought together the knowledge of physical scientists, engineers, industrialists, the military and politicians to synthesise basic science (such as achievement of controlled fission), the solution of a vast range of technical problems (such as developing an implosion trigger device), engineering and manufacturing prowess (as in generating adequate amounts of fissionable material) with military and political know-how (in the testing and deployment of the bombs) (Rhodes 1986).

The next class of methods—vision-based integration—involves having an ideal to work towards. The ideal is generally not tangible, like a product, but is a motivating force, which can lead to the development of a set of principles or some other unifying outcome. An example is the work of the World Commission on Dams, which has been guided by the idea of 'development effectiveness'—in other words, equitable and sustainable human development (World Commission on Dams 2000:xxxiii). The commission aimed to achieve a balance between demands for irrigation, electricity, flood control and water supply (the benefits of dams) and debt burden, displacement, the impoverishment of people and disturbance of ecosystems and fishery resources (the costs of dams). Consequently, in its research and consultation activities, the Commission worked with those displaced or otherwise affected by dams, as well as with powerful funders and construction industries—specifically, 'government agencies, project affected people and non-governmental organisations, people's movements, the dam construction industry, the export credit agencies and private investors, and the international development community' (World Commission on Dams 2000:viii). In particular, this involved synthesis of a range of technical, social, environmental, financial and economic evidence from case studies, country studies, a survey, technical reports, submissions and forums (see also Bammer 2006a).

Finally, common metrics-based methods involve converting various discipline and practice-based inputs into a single measure, such as a dollar value, global hectares of land, metric tonnes of carbon dioxide equivalent or disability-adjusted life years. Applying common metrics in research integration essentially has four steps: i) determining which common metric is most relevant to the research question in hand; ii) seeking disciplinary and stakeholder input to determine the parameters of the research problem that should be included in the common metric conversion and analysis; iii) where necessary, applying disciplinary and stakeholder knowledge to convert factors into the common metric value (for example, converting land use into a dollar value); and iv) combining the assigned values through simple arithmetic or some other manipulation (often modelling).

In terms of the third major group of methods—identifying and managing conflicts between, for example, values, interests and epistemologies—bringing a range of people with different perspectives together inevitably means that that there will be differences, and sometimes clashes, between them. The key integrative task is to maximise the insights from the different perspectives and to minimise unproductive conflicts and other such effects (Bammer 2008a). For example, principled negotiation, which focuses on differences in interests, is an effective tool for much dispute resolution. It concentrates on creative problem solving and fair accommodation of diverse interests (Bammer 2006b; Fisher et al. 1991; Gray 1989; Ury 1993). Identifying ways of dealing with other problematic differences is a fertile area for further research.

Finally, we suggest that a key problem with research integration is that there is no agreed standard way for describing or analysing it. This can lead to key elements being ignored, along with muddled thinking. Throughout this book, we use a simple framework based on six questions (Bammer & Land & Water Australia Integration Symposium Participants 2005).

1. What is the integration aiming to achieve and who is intended to benefit? This question aids thinking more clearly about the integrative purposes and differentiating them from other research aims, such as the development of new discipline-based knowledge.
2. What is being integrated? This helps considerations of the expertise that needs to be marshalled to achieve the integration aims. It also encourages clarity around the boundaries of the research, as discussed above.
3. Who is doing the integration? This question highlights that integration does not necessarily involve a group process. While the integrative process can be designed to involve everyone in the project, the task can be delegated to a subgroup or even one person. We suggest that an I2S specialist should be a key player.
4. How is the integration being undertaken? This question focuses attention on integrative methods. We suggest that there are five classes of integration methods, which we have outlined above.
5. What is the context for the integration? The question directs attention to the political or other action circumstances that have led to the research and which could be influential during its life. It also focuses on the institutions that are involved in funding or managing research integration. Integrated research is often undertaken in response to a driver from outside the research community, such as public concern, government policy or business innovation. Understanding the context can therefore be critical for appreciating how the research is shaped and the outcomes assessed.
6. What is the outcome of the integration? This involves examining what the integration produced, as well as the process of integration.

We use the framework throughout this book as a model for how such a systematic description could be structured. As we show in our examples, the elements can be presented in any order and can be combined when occasion demands.

Conclusion

In this appendix, we have briefly described a particular set of conceptual and method skills, which we call Integration and Implementation Sciences (I2S). We suggest that effectively tackling real-world problems requires a new type of researcher, who can draw together discipline and practice experts, and that such researchers need a solid foundation in the skills we outline here.

We propose that I2S can provide: a) the hub around which research institutions can organise teams to investigate real-world problems; b) a baseline level of quality for such work; c) a way of transmitting new ideas and methods between groups focusing on different real-world problems; and d) a home for drawing together and further developing recurrent issues in tackling real-world problems that are not within the domain of any discipline or practice area.

In terms of the last point, we suggest that I2S covers four domains—namely, concepts and methods to enhance:

* fresh thinking on intractable problems
* integration of disciplinary and stakeholder knowledge
* understanding and management of ignorance and uncertainty
* the provision of research support for decision making and practice change.

The dialogue methods we present here have been compiled as part of fleshing out the domain of the 'integration of disciplinary and stakeholder knowledge'. We have identified, along with dialogue-based methods, four other major groups of methods: model based, product based, vision based and common metric based. As well as these integration methods, the domain also includes concepts and methods to scope, frame and set boundaries around the real-world problem being investigated, as well as to identify and resolve inevitable conflicts around epistemologies, interests, values and so on. Systems-based thinking provides the core conceptual underpinning to this domain. Finally, we argue for an agreed standard way of describing and analysing research integration and provide a simple six-question framework, which we have found to be an effective starting point.

Appendix 2. Dialogue methods for research integration and the broader field of dialogue

While it is beyond the scope of this book to provide a summary of the broader field of dialogue, we do point interested readers to key references and show how our work relates to that field. There are three key aspects of the broader field. First is a literature that aims to develop theoretical foundations for dialogue more generally. Key references here are Bohm (1996), Isaacs (1999, 2001), Roberts (2002) and Yankelovich (1999). These references underpin Franco's approach to dialogue, which informs the work we present here. In particular, Franco (2006:814) points out:

> In contrast to debate and persuasion, participants in a dialogue do not attempt to validate particular propositions or find weaknesses in them. Rather, participants listen to find strength and value in another's position and work together towards a mutual understanding (Yankelovich 1999). According to Bohm (1996), the word 'dialogue' comes from the Greek 'dialogos': logos means 'the word' or the meaning of the word, and dia means 'through'. Dialogue involves the suspension of judgment or pre-conceptions, an equal participation in the conversation by the parties, empathetic listening, and the mutual probing of assumptions (Roberts, 2002). The goal of dialogue is to jointly create meaning and shared understanding between participants (Bohm, 1996; Isaacs, 1999, 2001; Yankelovich, 1999; Roberts, 2002).

Franco (2006), whose interest is in dialogue as a problem-structuring method, sees dialogue as one element in a typology of different conversational forms. He differentiates between debate, persuasion, dialogue, negotiation and deliberation. While these were useful distinctions to draw, we found that they were not completely applicable to our considerations of dialogue as a method for research integration. Certainly, we agree that a formal debate, strong advocacy and classical win–lose negotiation are unlikely to provide strong integrative tools. There are, however, forms of negotiation and deliberation that fit our definition of dialogue—namely, to 'jointly create meaning and shared understanding between participants' (Franco, 2006:814). In particular, we include principled negotiation, which involves each 'side' stating their interests and a formal process seeking a fair resolution of differences based on mutual respect and understanding. Furthermore, deliberation processes based on democratic principles can arguably also fit in a description of dialogue methods.

Dialogue can be used for multiple purposes, of which our use as a method for research integration is only one. These purposes can be wider than research and integration. In terms of the former, dialogue is useful in a diversity of contexts from setting policy to relationship counselling. In terms of the latter, we have already mentioned Franco's use of dialogue as a way to help groups systematically approach problems, as each group member will conceive the given problem differently. We provide two other examples. In Western intellectual thought, dialogue was first associated with the Greeks, particularly Socrates and Plato. For them, the key issue was 'reasoning through rigorous dialogue as a method for intellectual investigation intended to expose false beliefs and elicit truth' (Tarnas 1991:34). In the 1990s, Bohm (1996:vii) wrote about the use of dialogue as a process 'which explores an unusually wide range of human experience: our closely held values; the nature and intensity of emotions; the patterns of our thought processes; the function of memory; the import of inherited cultural myths; and the manner in which our neurophysiology structures moment-to-moment experience'. He went on to argue that '[i]n its deepest sense, then, dialogue is an invitation to test the viability of traditional definitions of what it means to be human, and collectively to explore the prospect of an enhanced humanity' (Bohm 1996:vii–viii). For those interested in dialogue more broadly, the references listed here provide some of the rich veins they can explore further.

A second key aspect of the broader field of dialogue is the literature on deliberative democracy. This is a theory or movement in political science, with deliberative democracy being defined as 'an association whose affairs are governed by the public deliberation of its members' (Cohen 1989:17) or, more broadly:

> Deliberative democracy is a conception of democratic politics in which citizens or their accountable representatives seek to give one another mutually acceptable reasons to justify the laws they adopt. The reasons are not merely procedural ('because the majority favours it') or purely substantive ('because it is a human right'). They appeal to moral principles (such as basic liberty or equal opportunity) that citizens who are motivated to find fair terms of cooperation can reasonably accept. (Gutmann and Thompson 2001:137)

Deliberative democracy can be contrasted with the more familiar approach of representative government in which democracy is realised through citizens voting for their elected representatives. Advocates of deliberative democracy point out that informed citizens, engaging in structured deliberations of important issues, produce a form of democracy that is more valid, and more constructive, than simply voting at elections. It is also differentiated from approaches that 'take fundamental rights as givens, and locate them as restraints on democratic decision-making (such as natural law conceptions and constitutionalism)' (Gutmann and Thompson 2001:137).

Structures and processes characterise deliberative democracy, and dialogue is central to both. Indeed, two of the dialogue methods discussed in this book—citizens' juries and the consensus conference—are detailed in *The Deliberative Democracy Handbook* (Gastil and Levine 2005). In many of the methods, lay people are presented with information and opinion by experts and key stakeholders in the domain under deliberation and, through facilitated discussion, integrate these inputs to develop a position (sometimes but not always a consensus position) to convey to decision makers.

A useful typology of methods is provided by Button and Ryfe (2005), who suggest that deliberative processes be classified by populating the cells of a 3x3 matrix, with the axes of the matrix being: a) who initiates the deliberative process (a civic association, a non-government organisation or a governmental organisation); and b) who participates (by self-selection, by random selection or by stakeholder selection). Thus, a neighbourhood association is an example of self-selection into a civic association, a deliberative poll is an example of random selection by a non-government organisation, a citizens' jury is an example of random selection by a government organisation, and deliberative locality planning is an example of stakeholder selection by a government organisation.

Rationality and impartiality are central to deliberative processes (Elster 1998) and all methods of deliberative democracy use dialogue to enhance both of these features. Rationality is enhanced through the combining of expert inputs and lay people's exercise of judgments, along with discussion to tease out and balance the arguments canvassed. Impartiality is enhanced as people make clear the basis of their understandings and judgments, through discussion test them against those of others and reflect on the similarities and differences in various players' positions.

The third key element of the broader field of dialogue is a range of compilations of dialogue methods and tool kits available. We list a number of these below and have discussed some in Chapter 2. We use many of these methods in this book, but we present them in a new way: as methods for research integration. As outlined above, we have looked at these methods to see if they are useful for integrating differences that are essential components of collaborative research.

We conclude this section by reiterating that not all dialogue requires a 'method'. The methods we present here are useful for structuring interactions and synthesis in groups that are large enough for the normal implicit rules of conversation to be less effective than when only two or a small number of people are involved. They can be useful in achieving broad understanding, as with the methods we describe for integrating judgments, as well as when synthesis of a particular aspect of difference in the research is required, as in integrating visions, interests, values or world views.

Tool kits that include dialogue methods

Canadian Rural Information Service, Agriculture and Agri-Food Canada 2002, *Community Dialogue Toolkit*, Canadian Rural Information Service, <http://www.rural.gc.ca/RURAL/>

Carson, L. and Gelber, K. 2001, *Ideas for Community Consultation: A discussion on principles and procedures for making consultation work*, NSW Department of Urban Affairs and Planning, Sydney, New South Wales.

Department of Sustainability and Environment, Victoria 2006, *Effective Engagement Toolkit*, <http://www.dse.vic.gov.au/DSE/wcmn203.nsf/childdocs/ -D15064A59E496FC8CA2570360014FEF8?open>

Keating, C. 2002, *Facilitation Toolkit: A practical guide for working more effectively with people and groups*, Department of Environment, Water and Catchment Protection, East Perth, Western Australia.

National Land and Water Resources Audit and ANZLIC—The Spatial Information Council 2003, *Natural Resources Information Management Toolkit*, National Land & Water Resources Audit, <http://www.nlwra.gov.au/ national-land-and-water-resources-audit/natural-resources-information-management-toolkit>

Start, D. and Hovland, I. 2004, *Tools for Policy Impact: A handbook for researchers*, Overseas Development Institute, London, <http://www.odi.org.uk/RAPID/Publications/ Tools_Policy_Impact.html>

Urban Research Program, Griffith University 2006, *URP Toolbox*, Urban Research Program, Griffith University (previously the *Coastal CRC's Citizen Science Toolbox*), <https://www3.secure.griffith.edu.au/03/toolbox/>

Appendix 3. How we developed this book

As we outlined in the introduction, our approach was iterative and switched between inductive and deductive. We cycled between identifying elements of research integration (such as synthesising facts, judgments, visions, values, interests, and so on), examining different types of dialogue and analysing case studies in order to match methods with integration tasks.

As we also described in the introduction, our work on identifying elements of research integration did not aim to be particularly comprehensive or precise. Instead, we sought to develop a list that was 'good enough' for interrogating the literature on dialogue and for hunting out case studies. In terms of the elements we identified—facts, judgments, visions, values, interests, epistemologies, time scales, geographical scales and world views—we did not find any dialogue methods geared specifically to integrating facts, epistemologies, time scales or geographical scales. The elements we list here are also not the ones we started with; instead, we started with a narrower band. As we read about dialogue methods, we reflected on what they could integrate and then expanded our list of research elements. The most noteworthy element that we added in this way was 'judgments'.

In terms of identifying and searching through available dialogue methods, colleagues helped us brainstorm a list and identify the various tool kits listed in Appendix 2. Again, this was added to in the course of the project. We started out very broadly, considering participatory as well as dialogue methods. In this way, we came up with 70 methods (see Appendix Table 3.1) from which we chose 14 to cover here. We ruled out participatory methods that did not involve dialogue, as well as those that did not seem to be useful for research integration. For example, focus groups are dialogue methods that are useful for gathering information, rather than for synthesis.

In terms of the search strategy for obtaining cases to illustrate the dialogue methods, our primary aim was to generate at least one good-quality case example for each dialogue method. Because we are interested in *research* integration, we have concentrated on academic articles. These have the added benefits of having been quality checked through peer review and of being readily obtainable through any large academic library. We based our search primarily on one electronic database: *Current Contents* (ISI—Thomson Scientific). This is a major, broad resource covering about 7500 journals across the sciences, social sciences and humanities. We concentrated on the period 1993 to the present. If there were few or no cases found, *Web of Science* (ISI—Thomson Scientific) was also searched, mainly via checking the citation listings of any descriptive or theoretical articles we had to hand to trace case articles. The Internet search engine *Google* was also used in these instances as this, on occasions, led to journal articles not listed in *Current Contents*, or to general resources through which case examples could be found.

A further restriction was that if an article describing a potential case study was not available electronically, it was passed over and another was selected. As the home organisation from which this search was conducted, The Australian National University, has extensive e-journal accessibility, this meant we could obtain rapidly most of what we identified as likely to be useful.

In terms of search practice, material on the dialogue method we were seeking to illustrate with cases was read and, from this, primary search terms were determined. Usually, this involved little more than taking the name of the dialogue method as the search term (for example, 'Delphi') with search controls to remove extraneous meanings. At times, however, more care had to be taken as some methods had multiple names or were difficult to separate through standard search strategies (for example, 'scenario planning' and 'scenario thinking'; 'search conferences' and 'future search conferences'). Within the initial search rotation, we focused on obtaining cases for each dialogue method from four target sectors: natural resource management (or, failing that, environment more generally), public health, security and technological innovation. For a number of the dialogue methods, we were not able to easily identify case examples in every sector. For a few, we found a good example in a different sector, which we decided to include. For principled negotiation, we were not able to identify any good case examples in *research* integration, so we chose one focusing on service provision. A summary of where we were able to find cases is provided in Table 2.1.

We present the best examples that we could find. For a few methods—for example, the Delphi technique—we were spoilt for choice. We found examples in each of our four areas of application and for various ways of combining discipline and stakeholder inputs, so that we could illustrate a range of ways of applying the method in research integration. In these instances, we sought cases that were complementary rather than identical. For example, if we had a case in natural resource management and we had two possible choices in security, we would use the security example that was most different from the case in natural resource management.

None of the dialogue methods we present is a tool solely for research integration. In other words, each method can also be used for purposes other than research in ways that do not involve researchers or that give them only a minor role. Because of our focus on research integration, we looked for examples where researchers had a role: in organising the dialogue, as facilitators, as participants, as 'expert witnesses' or in documenting the dialogue.

Most of the examples we found concentrated on stakeholder input. Examples where different disciplinary perspectives were brought together were less common, and illustrations combining disciplinary and stakeholder inputs were rare. That is not to say that the participants in dialogue for research integration always have to conform to a particular stereotype. On the contrary, the point we are making here is that the illustrations we are able to provide cover only a limited array of possibilities in terms of bringing various perspectives together.

Our aim in this book is not to be comprehensive, although we have included the majority of methods for which we could find reasonable descriptions. Primarily, we wanted to get a sense of the array of available methods and to explore how well they linked with specific research integration tasks. As far as we are aware, this has not been done previously. We did not include methods that were not yet well documented, even though a number of these were drawn to our attention. We urge colleagues to write these up. We are keen to hear about documented methods we have missed.

Appendix Table 3.1 Methods reviewed for their usefulness as dialogue methods for research integration

Action learning

Action research

Appreciative inquiry

Backcasting

Battlemap

Boston box

Charrette

Citizen committees

Citizens' juries

Community fairs

Community indicator

Consensus conference

Consensus development panel

Copy platform

Critical systems heuristics

Deliberative dialogue

Deliberative forums

Deliberative polling

Delphi technique

Electronic democracy

Engaging public participation

Episode studies

Ethical matrix

Expert panel

Field trips

Fishbowl

Focus group discussion

Force field analysis

Future search conference

Interactive TV

Interactive video display kiosks

Kitchen table discussion

Lens workshop

Market segmentation

Marketing approach

Marketing matrix

Most significant change technique

Multi-criteria decision analysis

Multi-objective decision-making support

Nominal group technique

Open space technology

Organisation readiness assessment

Participatory development

Participatory (rural) appraisal

Photovoice

Planning4real

Positioning

Principled negotiation

Prioritisation matrix

Problem tree analysis

Promotions matrix

Public conversation

Public involvement volunteers

Public meetings

Rapid assessment

Residents' feedback panel

Role-plays

Samoan circles

Scenario planning

Search conference

Sketch interviews

Social learning

Sociotechnical systems thinking

Soft systems methodology

Speakouts

Strategic assumption surfacing and testing

Strategic planning

Study circles

SWOT analysis

Triangle analysis

Visioning

Writeshops

References

Adler, M. and Ziglio, E. (eds) 1996, *Gazing into the Oracle: The Delphi method and its application to social policy and public health*, Jessica Kingsley Publishers, London.

Aldred, J. and Jacobs, M. 2000, 'Citizens and wetlands: evaluating the Ely citizens' jury', *Ecological Economics*, vol. 34, no. 2, pp. 217–32.

Appreciative Inquiry Commons n.d., *Appreciative Inquiry Commons*, Weatherhead School of Management at Case Western Reserve University, <http://appreciativeinquiry.case.edu/>

Balkany, T. J., Hodges, A. V., Buchman, C. A., Luxford, W. M., Pillsbury, C. H., Roland, P. S., Shallop, J. K., Backous, D. D., Franz, D., Graham, J. M., Hirsch, B., Luntz, M., Niparko, J. K., Patrick, J., Payne, S. L., Telischi, F. F., Tobey, E. A., Truy, E. and Staller, S. 2005, 'Cochlear implant soft failures: consensus development conference statement', *Otology and Neurotology*, vol. 26, no. 4, pp. 815–18.

Bammer, G. 2005, 'Integration and implementation sciences: building a new specialization', *Ecology and Society*, vol. 10, no. 2, article 6, <http://www.ecologyandsociety.org/vol10/iss2/art6/>.

Bammer, G. 2006a, Illustrating a systematic approach to explain integration in research—The case of the World Commission on Dams, *Integration Insights*, no. 2, October <http://www.anu.edu.au/iisn/activities/integration_insights/integration-insight_2.pdf>.

Bammer, G. 2006b, 'Principled negotiation—a method for integrating interests', *Integration Insights*, no. 3, <http://www.anu.edu.au/iisn/activities/integration_insights/integration-insight_3.pdf>.

Bammer, G. 2006c, 'Scoping public health problems', in D. Pencheon, C. Guest, D. Melzer and J. A. M. Gray (eds), *Oxford Handbook of Public Health Practice*, Second edition, Oxford University Press, Oxford, pp. 4–11.

Bammer, G. 2006d, *A systematic approach to integration in research, Integration Insights*, no. 1, <http://www.anu.edu.au/iisn/activities/integration_insights/integration-insight_1.pdf>.

Bammer, G. 2008a, 'Enhancing research collaboration: three key management challenges', *Research Policy*, vol. 37, no. 5, pp. 875–87.

Bammer, G. 2008b 'Do we need a new discipline to document and transmit problem-based learnings?' *Proceedings Integrated Design and Process Technology IDPT, June 1-6, Taichung, Taiwan (annual conference of the Society for Design and Process Science)*.

Bammer, G. & Land & Water Australia Integration Symposium Participants 2005, 'Guiding principles for integration in natural resource management (NRM) as a contribution to sustainability', *Australasian Journal of Environmental Management*, vol. 12, Supplementary issue, pp. 5–7.

Bammer, G. and Smithson, M. (eds) 2008, *Uncertainty and Risk: Multidisciplinary perspectives*, Earthscan, London.

Beauchamp, T. L. and Childress, J. F. 2001, *Principles of Biomedical Ethics*, Fifth edition, Oxford University Press, Oxford.

Bissell, P., Ward, P. R. and Noyce, P. R. 2000, 'Appropriateness measurement: application to advice-giving in community pharmacies', *Social Science & Medicine*, vol. 51, no. 3, pp. 343–59.

Bohm, D. 1996, *On Dialogue*, ed. L. Nichol, Routledge, London and New York.

Bradfield, R., Wright, G., Burt, G., Cairns, G. and Van Der Heijden, K. 2005, 'The origins and evolution of scenario techniques in long range business planning', *Futures*, vol. 37, no. 8, pp. 795–812.

Brocklehurst, N. J., Hook, G., Bond, M. and Goodwin, S. 2005, 'Developing the public health practitioner work force in England: lessons from theory and practice', *Public Health*, vol. 119, no. 11, pp. 995–1002.

Bryson, J. M. and Anderson, S. R. 2000, 'Applying large-group interaction methods in the planning and implementation of major change efforts', *Public Administration Review*, vol. 60, no. 2, pp. 143–62.

Button, M. and Ryfe, D. M. 2005, 'What can we learn from the practice of deliberative democracy?', in J. Gastil and P. Levine (eds), *The Deliberative Democracy Handbook: Strategies for effective civic engagement in the twenty-first century*, Jossey-Bass, San Francisco, pp. 20–33.

Caldwell, R. L. n.d., *Scenarios, Foresight and Change. Tutorial 2: Building scenarios*, University of Arizona, <http://ag.arizona.edu/futures/tou/tut2-buildscenarios.html>.

Canadian Rural Information Service, Agriculture and Agri-Food Canada 2002, *Community Dialogue Toolkit*, Canadian Rural Information Service, <http://www.rural.gc.ca/RURAL>.

Carson, L. and Gelber, K. 2001, *Ideas for Community Consultation: A discussion on principles and procedures for making consultation work*, NSW Department of Urban Affairs and Planning, Sydney, New South Wales.

Carson, L., Sargant, C. and Blackadder, J. 2004, *Consult Your Community: A guide to running a youth jury*, NSW Premier's Department, Sydney, New South Wales.

Checkland, P. 1981, *Systems Thinking, Systems Practice*, J. Wiley, Chichester, Sussex.

Checkland, P. 1985, 'From optimizing to learning: a development of systems thinking for the 1990s', *Journal of the Operational Research Society*, vol. 36, no. 9, pp. 757–67.

Checkland, P. and Poulter, J. 2006, *Learning for Action: A short definitive account of soft systems methodology and its use for practitioner, teachers, and students*, John Wiley and Sons, Hoboken, NJ.

Checkland, P. and Scholes, J. 1999, *Soft Systems Methodology in Action*, New edition, Wiley, New York.

City of Greater Bendigo n.d., Bendigo+25 Future Search Conference, City of Greater Bendigo, <http://www.bendigo.vic.gov.au/Page/Page.asp?Page_Id=2016&h=1>.

Cohen, J. 1989, 'Deliberation and democratic legitimacy', in A. P. Hamlin and P. Pettit (eds), *The Good Polity: Normative analysis of the state*, Blackwell, Oxford, pp. 17–34.

Coico, R., Kachur, E., Lima, V. and Lipper, S. 2004, 'Guidelines for preclerkship bioterrorism curricula', *Academic Medicine*, vol. 79, no. 4, pp. 366–75.

Cooperrider, D. L. and Whitney, D. K. 2005, *Appreciative Inquiry: A positive revolution in change*, Berrett-Koehler, San Francisco, Calif.

Coote, A. and Lenaghan, J. 1997, *Citizens' Juries: Theory into practice*, Institute for Public Policy Research, London.

Corbin, J. M. and Strauss, A. L. 2008, *Basics of Qualitative Research: Techniques and procedures for developing grounded theory*, Third edition, Sage Publications, Inc., Thousand Oaks, Calif.

Cousin, G. and Healey, M. 2003, 'Pedagogic research methods in geography higher education', *Journal of Geography in Higher Education*, vol. 27, no. 3, p. 341-7.

Crosby, N. and Nethercut, D. 2005, 'Citizens juries: creating a trustworthy voice of the people', in J. Gastil and P. Levine (eds), *The Deliberative Democracy Handbook: Strategies for effective civic engagement in the twenty-first century*, Jossey-Bass, San Francisco, pp. 111–19.

Dalkey, N. and Helmer, O. 1963, 'An experimental application of the Delphi method to the use of experts', *Management Science*, vol. 9, no. 3, pp. 458–67.

Dart, J. and Davies, R. 2003, 'A dialogical, story-based evaluation tool: the most significant change technique', *American Journal of Evaluation*, vol. 24, no. 2, pp. 137–55.

Dash, D. 2007, *SAST Methodology*, Xavier Institute of Management, Bhubaneswar, India,
<http://www1.ximb.ac.in/users/fac/dpdash/dpdash.nsf/pages/CP_SAST>.

Davies, R. 1996, *An Evolutionary Approach to Facilitating Organisational Learning: An experiment by the Christian Commission for Development in Bangladesh*, Centre for Development Studies, Swansea, Wales.

Davies, R. and Dart, J. 2005, *The Most Significant Change (MSC) Technique: A guide to its use*, Rick Davies and Jess Dart, Trumpington, Cambridge, United Kingdom, and Hastings, Victoria, Australia.

De Jouvenel, H. 2000, 'A brief methodological guide to scenario building', *Technological Forecasting and Social Change*, vol. 65, no. 1, pp. 37–48.

Delbecq, A. L., Gustafson, D. H. and Van de Ven, A. H. 1975, *Group Techniques for Program Planning: A guide to nominal group and Delphi processes*, Management Application Series, Scott, Foresman, Glenview, Ill.

Department of Sustainability and Environment, Victoria 2006, *Effective Engagement Toolkit*,
<http://www.dse.vic.gov.au/DSE/wcmn203.nsf/childdocs/ -D15064A59E496FC8CA2570360014FEF8?open>.

Dunham, R. B. 1998, *Nominal Group Technique: A users' guide*,
<http://courses.bus.wisc.edu/rdunham/EMBA/Fall_2006_Readings/ TEAMS/dunham_ngt.pdf>.

Elster, J. (ed.) 1998, *Deliberative Democracy*, Cambridge Studies in the Theory of Democracy, Cambridge University Press, London.

Emery, M. and Purser, R. E. 1996, *The Search Conference: A powerful method for planning organizational change and community action*, Jossey-Bass Public Administration Series, Jossey-Bass, San Francisco, Calif.

Fahey, L. and Randall, R. M. (eds) 1998, *Learning From the Future: Competitive foresight scenarios*, Wiley, New York.

Fisher, R. and Shapiro, D. 2005, *Beyond Reason: Using emotions as you negotiate*, Viking, New York.

Fisher, R., Ury, W. and Patton, B. 1991, *Getting to Yes: Negotiating an agreement without giving in*, Second edition, Random House, London.

Flood, R. L. and Jackson, M. C. 1991, *Creative Problem Solving: Total systems intervention*, Wiley, Chichester, New York.

Food Ethics Council 2001, *After FMD: Aiming for a values-driven agriculture*, Food Ethics Council, Brighton, United Kingdom.

Food Ethics Council n.d., *Ethical Matrix*, Food Ethics Council,
<http://www.foodethicscouncil.org/ourwork/tools/ethicalmatrix/introduction>.

Franco, L. A. 2006, 'Forms of conversation and problem structuring methods: a conceptual development', *Journal of the Operational Research Society*, vol. 57, pp. 813–21.

Fuller, I., Gaskin, S. and Scott, I. 2003, 'Student perceptions of geography and environmental science fieldwork in the light of restricted access to the field, caused by foot and mouth disease in the UK in 2001', *Journal of Geography in Higher Education*, vol. 27, no. 1, pp. 79–102.

Future Search Network 2003, *Future Search Network*, <http://www.futuresearch.net/>.

Gastil, J. and Levine, P. (eds) 2005, *The Deliberative Democracy Handbook: Strategies for effective civic engagement in the twenty-first century*, First edition, Jossey-Bass, San Francisco.

Gray, B. 1989, *Collaborating: Finding common ground for multiparty problems*, Jossey-Bass, San Francisco.

Gregory, W. J. and Midgley, G. 2000, 'Planning for disaster: developing a multi-agency counselling service', *Journal of the Operational Research Society*, vol. 51, no. 3, pp. 278–90.

Grundahl, J. 1995, 'The Danish consensus conference model', in S. Joss and J. Durant (eds), *Public Participation in Science: The role of consensus conferences in Europe*, Science Museum with the support of the European Commission Directorate General XII, London, pp. 31–40.

Guston, D. H. 1999, 'Evaluating the first US consensus conference: the impact of the citizens' panel on telecommunications and the future of democracy', *Science Technology & Human Values*, vol. 24, no. 4, pp. 451–82.

Gutmann, A. and Thompson, D. 2001, 'Deliberative democracy', in P. A. B. Clarke and J. Foweraker (eds), *Encyclopedia of Democratic Thought*, Routledge, London, pp. 137–41.

Hammond, S. A. 1998, *The Thin Book of Appreciative Inquiry*, Second edition, Thin Book Publishing Company, Plano, Tex.

Hammond, S. A. and Royal, C. (eds) 1998, *Lessons From the Field: Applying appreciative inquiry*, Practical Press, Plano, Tex.

Heft, L. n.d., *A Description of Open Space Technology*, <http://www.openingspace.net/ openSpaceTechnology_method_DescriptionOpenSpaceTechnology.shtml>.

Hendriks, C. 2005, 'Consensus conferences and planning cells', in J. Gastil and P. Levine (eds), *The Deliberative Democracy Handbook: Strategies for effective civic engagement in the twenty-first century*, Jossey-Bass, San Francisco, pp. 80–110.

Henson, S. 1997, 'Estimating the incidence of food-borne *Salmonella* and the effectiveness of alternative control measures using the Delphi method', *International Journal of Food Microbiology*, vol. 35, no. 3, pp. 195–204.

Herkert, J. R. and Nielsen, C. S. 1998, 'Assessing the impact of shift to electronic communication and information dissemination by a professional organization: an analysis of the Institute of Electrical and Electronics Engineers (IEEE)', *Technological Forecasting & Social Change*, vol. 57, no. 1–2, pp. 75–103.

Holwell, S. 2000, 'Soft systems methodology: other voices', *Systemic Practice & Action Research*, vol. 13, no. 6, pp. 773–97.

Isaacs, W. 1999, *Dialogue and the Art of Thinking Together. A pioneering approach to communicating in business and in life*, Currency/Doubleday, New York.

Isaacs, W. 2001, 'Towards an action theory of dialogue', *International Journal of Public Administration*, vol. 24, pp. 709–48.

Jefferson Center 2004, *Citizens Jury Handbook*, Revised and updated edition, The Jefferson Center, <http://www.jefferson-center.org/>.

Joss, S. and Durant, J. (eds) 1995, *Public Participation in Science: The role of consensus conferences in Europe*, Science Museum with the support of the European Commission Directorate General XII, London.

Kaiser, M. and Forsberg, E.-M. 2001, 'Assessing fisheries—using an ethical matrix in a participatory process', *Journal of Agricultural & Environmental Ethics*, vol. 14, no. 2, pp. 191–200.

Kashefi, E. and Mort, M. 2004, 'Grounded citizens' juries: a tool for health activism?', *Health Expectations*, vol. 7, no. 4, pp. 290–302.

Keating, C. 2002, *Facilitation Toolkit: A practical guide for working more effectively with people and groups*, Department of Environment, Water and Catchment Protection, East Perth, Western Australia.

Kueffer, C., Hadorn, G. H., Bammer, G., van Kerkhoff, L. and Pohl, C. 2007, 'Towards a publication culture in transdisciplinary research', *GAIA*, vol. 16, no. 1, pp. 22–6.

Lens, V. 2004, 'Principled negotiation: a new tool for case advocacy', *Social Work*, vol. 49, no. 3, pp. 506–13.

Linstone, H. A. and Turoff, M. (eds) 1975, *The Delphi Method: Techniques and applications*, Addison-Wesley Publishing Company Advanced Book Program, Reading, Mass.

Lockwood, M. 2005, 'Integration of natural area values: conceptual foundations and methodological approaches', *Australasian Journal of Environmental Management*, vol. 12 supp., pp. 8–19.

Maani, K. E. and Cavana, R. Y. 2007, *Systems Thinking, System Dynamics: Managing change and complexity*, Pearson Education New Zealand, North Shore, New Zealand.

Mackey, J. 2008, 'Musical improvisation, creativity and uncertainty', in G. Bammer and M. Smithson (eds), *Uncertainty and Risk: Multidisciplinary perspectives*, Earthscan, London, pp. 105–13.

Macquarie Dictionary 2005, *WordGenius 4.2*, October 2005 (W32), Revised third edition, Eurofield Information Solutions, Chatswood, New South Wales, Australia, <www.wordgenius.com.au>.

Mason, R. O. and Mitroff, I. I. 1981, *Challenging Strategic Planning Assumptions: Theory, cases, and techniques*, Wiley, New York.

McDonald, D., Bammer, G. and Breen, G. 2005, *Australian illicit drugs policy: mapping structures and processes*, Drug Policy Modelling Project Monograph Series No. 4, Turning Point Alcohol and Drug Centre, Fitzroy, Victoria, <http://www.dpmp.unsw.edu.au/DPMPWeb.nsf/ resources/DPMP+Monographs1/$file/DPMP+MONO+4.pdf>.

Mepham, B. 2000, 'A framework for the ethical analysis of novel foods: the ethical matrix', *Journal of Agricultural and Environmental Ethics*, vol. 12, no. 2, pp. 165–76.

Mepham, B., Kaiser, M., Thorstensen, E., Tomkins, S. and Millar, K. 2006, *Ethical Matrix Manual*, Agricultural Economics Research Institute (LEI), The Hague, <http://www.ethicaltools.info/content/ET2 Manual EM (Binnenwerk 45p).pdf>.

Midgley, G. 2000, *Systemic Intervention: Philosophy, methodology, and practice*, Kluwer Academic/Plenum Publishers, New York.

Midgley, G. (ed.) 2003, *Systems Thinking*, 4 vols, Sage, London.

Millar, K., Thorstensen, E., Tomkins, S., Mepham, B. and Kaiser, M. 2007, 'Developing the ethical Delphi', *Journal of Agricultural and Environmental Ethics*, vol. 20, no. 1, pp. 53–63.

Millar, K., Tomkins, S., Thorstensen, E., Mepham, B. and Kaiser, M. 2006, *Ethical Delphi Manual*, Agricultural Economics Research Institute (LEI), The Hague, <http://www.ethicaltools.info/content/ET3 Manual ED (Binnenwerk 43p).pdf>.

Mitroff, I. I. and Turoff, M. 1975, 'Philosophical and methodological foundations of Delphi', in H. A. Linstone and M. Turoff (eds), *The Delphi Method:*

Techniques and applications, Addison-Wesley Publishing Company Advanced Book Program, Reading, Mass., pp. 17–36.

Moyer, K. 1996, 'Scenario planning at British Airways—a case study', *Long Range Planning*, vol. 29, no. 2, pp. 172–81.

National Institutes of Health (NIH) 2006, 'NIH State-of-the-Science Conference statement on tobacco use: prevention, cessation, and control', 2006, *NIH Consensus and State-of-the-Science Statements*, vol. 23, no. 3, pp. 1–26.

National Institutes of Health (NIH) 2008, *Consensus Development Program: About us*, Consensus Development Program, National Institutes of Health, <http://consensus.nih.gov/ABOUTCDP.htm>.

National Institutes of Health State-of-the-Science Panel 2006, 'National Institutes of Health State-of-the-Science Conference statement: tobacco use: prevention, cessation, and control', *Annals of Internal Medicine*, vol. 145, no. 11, pp. 839–44.

National Land and Water Resources Audit and ANZLIC—The Spatial Information Council 2003, *Natural Resources Information Management Toolkit*, National Land & Water Resources Audit, <http://www.nlwra.gov.au/national-land-and-water-resources-audit/natural-resources-information-management-toolkit>.

Oels, A. 2002, 'Investigating the emotional roller-coaster ride: a case study-based assessment of the future search conference design', *Systems Research & Behavioral Science*, vol. 19, no. 4, pp. 347–55.

Owen, H. 1997a, *Expanding Our Now: The story of open space technology*, Berrett-Koehler Publishers, San Francisco.

Owen, H. 1997b, *Open Space Technology: A user's guide*, Second edition, Berrett-Koehler Publishers, San Francisco.

Owen, H. n.d., *A Brief Users' Guide to Open Space Technology*, <http://www.openspaceworld.com/users_guide.htm>.

Owen, H. (ed.) 1995, *Tales From Open Space*, Abbott Publishers, Potomac, Md.

Padgett, D. A. and Imani, N. O. 1999, 'Qualitative and quantitative assessment of land-use managers' attitudes toward environmental justice', *Environmental Management*, vol. 24, no. 4, pp. 509–15.

Penker, M. and Wytrzens, H. K. 2005, 'Scenarios for the Austrian food chain in 2020 and its landscape impacts', *Landscape & Urban Planning*, vol. 71, no. 2–4, pp. 175–89.

Pfeiffer, J. W. and Jones, J. E. 1975, *Reference Guide to Handbooks and Annuals*, University Associates, La Jolla, Calif.

Polanyi, M. 2001, 'Toward common ground and action on repetitive strain injuries: an assessment of a future search conference', *Journal of Applied Behavioral Science*, vol. 37, no. 4, pp. 465–87.

Ranney, L., Melvin, C., Lux, L., McClain, E., Morgan, L. and Lohr, K. N. 2006, 'Tobacco use: prevention, cessation, and control', *Evidence Report — Technology Assessment (Full Report)*, no. 140, pp. 1–120.

Reed, J., Pearson, P., Douglas, B., Swinburne, S. and Wilding, H. 2002, 'Going home from hospital—an appreciative inquiry study', *Health and Social Care in the Community*, vol. 10, no. 1, pp. 36–45.

Rhodes, R. 1986, *The Making of the Atomic Bomb*, Simon & Schuster, New York.

Roberts, N. 2002, *Transformative Power of Dialogue*, Elsevier, London.

Schroeder, D. and Palmer, C. 2003, 'Technology assessment and the "ethical matrix"', *Poiesis & Praxis: International Journal of Technology Assessment and Ethics of Science*, vol. 1, no. 4, pp. 295–307.

Srbljinovic, A. and Skunca, O. 2003, 'Agent based modelling and simulation of social processes', *Interdisciplinary Description of Complex Systems*, vol. 1, no. 1–2, pp. 1–8.

Start, D. and Hovland, I. 2004, *Tools for Policy Impact: A handbook for researchers*, Overseas Development Institute, London.

Tarnas, R. 1991, *The Passion of the Western Mind. Understanding the ideas that have shaped our world view*, Pimlico (Random House), London.

The Oxford English Dictionary 1989, 'value, n. 6.a.', *The Oxford English Dictionary*, Second edition, OED Online, Oxford University Press.

Trochim, M. K. and Trochim, W. M. K. 2007, 'Concept mapping for planning and evaluation', *Applied Social Research Methods*, no. 50, Sage Publications, Thousand Oaks, Calif.

Urban Research Program, Griffith University 2006, *URP Toolbox*, Urban Research Program, Griffith University, <https://www3.secure.griffith.edu.au/03/toolbox>.

Ury, W. L. 1993, *Getting Past No: Negotiating your way from confrontation to cooperation*, Bantam Books, New York.

van Vuuren, L. J. and Crous, F. 2005, 'Utilising appreciative inquiry (AI) in creating a shared meaning of ethics in organisations', *Journal of Business Ethics*, vol. 57, pp. 399–412.

Wakeford, T. 2002, 'Citizens juries: a radical alternative for social research', *Social Research Update*, no. 37, <http://sru.soc.surrey.ac.uk/SRU37.html>.

Watkins, J. M. and Mohr, B. J. 2001, *Appreciative Inquiry: Change at the speed of imagination*, Practicing Organization Development Series, Jossey-Bass/Pfeiffer, San Francisco, Calif.

Weisbord, M. R. (ed.) 1992, *Discovering Common Ground: How future search conferences bring people together to achieve breakthrough innovation, empowerment, shared vision, and collaborative action*, Berrett-Koehler, San Francisco.

Weisbord, M. R. and Janoff, S. 2000, *Future Search: An action guide to finding common ground in organizations and communities*, Second edition, Berrett-Koehler, San Francisco.

Wikipedia Contributors 2009, 'Evolutionary epistemology', *Wikipedia. The Free Encyclopedia*, <http://en.wikipedia.org/w/index.php?title=Evolutionary_epistemology&oldid=270028109>.

Wilson, I. 2000, 'From scenario thinking to strategic action', *Technological Forecasting and Social Change*, vol. 65, no. 1, pp. 23–9.

World Commission on Dams 2000, *Dams and development: a new framework for decision-making*, The Report of the World Commission on Dams, Earthscan, London.

Wright, T. S. A. 2006, 'Giving "teeth" to an environmental policy: a Delphi study at Dalhousie University', *Journal of Cleaner Production*, vol. 14, no. 9–11, pp. 761–8.

Yankelovich, D. 1999, *The Magic of Dialogue. Transforming conflict into cooperation*, Simon & Schuster, New York.

Index

www.ingramcontent.com/pod-product-compliance
Lightning Source LLC
Chambersburg PA
CBHW061246270326
41928CB00041B/3443